Engineering the Guitar

Theory and Practice

T0214923

Richard Mark French

Engineering the Guitar

Theory and Practice

Springer

Richard Mark French
Purdue University
Department of Mechanical Engineering Technology
Knoy Hall of Technology
West Lafayette, IN 47907-2021

ISBN: 978-1-4419-4496-2   e-ISBN: 978-0-387-74369-1
DOI: 10.1007/978-0-387-74369-1

I dedicate this book to my wife, Amy

Preface

There are some good books on the market that teach the craft of making guitars. Many are well-illustrated and can be very good sources of information for the beginning guitar maker. There is, in addition, a smaller number of books on the market dealing with the science of musical instruments and guitars in particular. This book is intended to bridge the gap between the craftsman and the scientist by presenting the technical considerations involved with designing and manufacturing guitars. With this in mind, I have written for an audience that wants to know how to engineer a guitar and is not put off by the mathematics necessary to describe the various topics.

I am an assistant professor in a mechanical engineering technology department and I have, perhaps predictably, written with my own students in mind. They are mathematically literate, but not always comfortable with calculus. Thus, I have avoided calculus where I didn't think it was required. The topics I have included and the ways in which I have presented them have also been strongly influenced by people working in the guitar industry. There is currently no comprehensive, practice-oriented work to which they can refer and I hope to at least partially address their needs as well.

I have tried to include practical information in the text based on my own experience and that of others who have offered their suggestions. This includes tables listing nominal dimensions of different types of instruments and basic properties of components such as tuners and pickups. I have also tried to describe current best practices in guitar design and manufacturing and to identify promising developments that may advance the field.

Much of the best work in the field of musical acoustics has been done by physicists or at least written with them in mind. For an excellent overview, I encourage you read *The Physics of Musical Instruments* by Neville Fletcher and Thomas Rossing. For an equally outstanding overview of acoustics including the basis of musical scales, I encourage you to read *The Science of Sound* by Thomas Rossing, Richard Moore and Paul Wheeler. I keep them both close at hand and continue to learn from both.

Since I have written this book for the technically inclined builder, I fear that it may seem quirky or even muddled to someone with a background in the physical sciences rather than engineering or manufacturing. There are many points in the text where I have been quite conscious of violating what such a reader might rightly feel are basic rules of technical writing. For example, I have generally presented measurements in decimal inches as well as millimeters. There is clearly no place in scientific communications for the English system of units, but American guitar manufacturers generally work in decimal inches. In order to be relevant to them, I feel compelled to use their conventions.

Another potential source of confusion for a reader not familiar with manufacturing processes is the presentation of dimensions to the nearest 0.001 inches. While a builder using simple fixtures or hand processes is very unlikely to work to this precision, computer-controlled machines are generally capable of it. As long as the temperature and humidity are carefully controlled, it is reasonable to work to a few 1/1000ths of an inch using the proper equipment.

One section which might seem a little long is the section in chapter 3 on calculating the deformation of the neck under string forces. There are relatively few parts of a guitar for which static or dynamic deflections can be calculated using simple analytical methods. The neck is one of these and this section is perhaps more detailed than one would expect so that it can be used as a teaching tool for students who have taken a basic course in strength of materials.

Finally, I wanted very much to produce a well-illustrated book and was fortunate to find researchers and instrument makers willing to supply high resolution digital images. In the hopes that future editions of this or some other book might further contribute to the store of literature available to technically-inclined luthiers, I would welcome other images that illustrate some potentially interesting aspect of the field.

Mark French
West Lafayette, Indiana 2008
rmfrench@purdue.edu

Acknowledgements

Nearly every book has a section at the front in which the author thanks the people who were instrumental in bringing it into being. Over the years, I've read these sections (one can't claim to have read a book without having read the whole thing, right?) and wondered how it could possibly take so many people to create a book. Well, now I know. I offer my thanks to the following people; certainly, this book would never have come to be without them.

First, many thanks are due to the people who have so generously shared their knowledge and experience in making guitars. My first contacts in the guitar industry were Mike Voltz at Gibson and Bill Hudak, then at Ovation. Both were kind and patient with me when I was new to the field. The whole crew at Taylor Guitars has been wonderful. Bob Taylor has set the standard for opening his factory to visitors. He has said he wants to leave the guitar in a better state than he found it; certainly he has succeeded. Dave Hosler, a former circus flyer who, through some amazing succession of events, has become one of the sources of innovation at Taylor has been gracious, encouraging and open at all turns. Thanks also to Brian Swerdfeger, David Judd, Ed Granero, Matt Guzetta and the rest of the gang there.

I have also been fortunate enough to work with Tim Shaw and Josh Hurst from Fender Guitars. Tim has about as much guitar industry experience as exists in a single human. He's friendly, encouraging and generous with his knowledge; I have benefited greatly from him. Josh has been helpful, enthusiastic and a great source of information on how modern design tools are changing the industry.

Kevin Beller at Seymour Duncan has been gracious and supportive. He also took time he probably couldn't really spare to review parts of this book.

I owe special thanks to Gene Maddux, a friend, colleague and mentor who first suggested to me that the dynamics of guitars might be an interesting thing to study. He not only got me thinking about the subject, but also let me use his photomechanics lab to study the behavior of a number of instruments. His patience and guidance changed the direction of my career at a time when I had thoughts of giving it up. May every young engineer have such an influence.

Thanks are also due to the faculty and staff at Purdue University for the freedom to do this work and for their continuing support. There are too few places left in academia for work that isn't clearly directed towards large funding agencies. Additionally, I wish to thank Mike Jacob in the Department of Electrical and Computer Engineering Technology for taking time to remind this aerospace engineer how to do basic circuit analysis.

x   Acknowledgements

Much of the help I received was of a non-technical nature. Elaine Tham and Lauren Danahy at Springer were helpful and encouraging during the long writing process. I am grateful to Elaine for taking a chance on someone who had never written a book before. Kay Solomon, undergraduate advisor in the Department of Mechanical Engineering Technology at Purdue and part-time literary goddess, graciously read several drafts of the manuscript and made many helpful suggestions. She is a bright spot in many lives.

Thanks are due also to my anonymous reviewers. They read several drafts of the manuscript and offered many good suggestions. It's never fun to subject one's writing to anonymous review, but the result is well worth the discomfort inflicted on the ego. One of the reviewers, in particular, went far beyond the call of duty by providing hand-written comments on complete drafts of the manuscript. Reviewers provide a necessary an often un-noticed service to the larger technical community; more than once, mine saved me from myself. Any errors remaining in the manuscript are my own.

Finally, completing a project like this one would be much more difficult without the support of a loving family. I am most fortunate to have a large, loud, gregarious, supportive family around me. Brian and Kate are a continuing source of joy and are remarkably tolerant of a Dad who wants to write books. Most of all, I owe many thanks to my lovely wife, Amy. She has encouraged me, pushed me when I needed it and seems not to mind having a distracted professor for a husband. I am a lucky man indeed.

Contents

1 History of the Guitar

Stringed instruments almost certainly predate recorded history. Anthropologists believe that the bow and arrow was perhaps the first machine made by early man and stringed instruments may have followed shortly afterwards [1, 2]. Stringed instruments have appeared in various forms and in many different cultures throughout recorded history. However, the story of the guitar is a recent one.

1.1 Development of the Classical Guitar

Early versions of the guitar appeared at least as early as the Baroque period (roughly 1600–1750) and Antonio Stradivari is known to have made guitars [3]. Figure 1.1 shows a modern reproduction of a baroque guitar with a 10 strings arranged in five pairs called courses.

The classical guitar really began to take its modern form in the early 19th century. By then, the tradition of stringing instruments in courses had given way to a guitar with six single strings and the modern E-A-D-G-B-E tuning. In 1800 the popularity of the guitar was still concentrated in Spain and Italy where it was already taken seriously as an instrument. The body of music available for it was growing, but still hadn't reached critical mass. However, this was about to change with the emergence of a group of gifted guitarists who also composed music for the instrument [4].

Fernando Sor (1778–1839) was perhaps the greatest of this small group that helped transform the guitar into an instrument for serious musicians all over Europe. Sor was born in Barcelona and moved to Paris in 1813, where he earned a reputation as a master performer and composer. His reputation remains intact as noted classical guitarists still record his compositions and they are still found among current collections of sheet music for students of the classical guitar.

As the guitar evolved through the early 1800s, geared tuners began to replace simple friction pegs. The only modern guitars that still use tuning pegs are flamenco guitars. In addition to the introduction of geared tuners, the flush mounted fretboard was replaced by one that covered the neck as well as part of the soundboard.

Though there were many guitar builders around Europe by the mid 19th century, the one who generally gets the credit for being the father of the modern classical guitar is Antonio Torres (1817–1892) [4-5]. His instruments incorporated incremental improvements over those of his predecessors and included design features that had been introduced by other luthiers. This evolutionary process is hardly a surprise, since almost all new inventions grow from an existing technological base. The resulting classical guitar was a refined instrument suitable for demanding players and composers.

R.M. French, *Engineering the Guitar*,
DOI: 10.1007/978-0-387-74369-1_1, © Springer Science+Business Media, LLC 2009

Figure 1.1 Modern Reproduction of a Baroque Guitar (image courtesy of Daniel Larson, www.daniellarson.com)

Torres' instruments had wider bodies, essentially the modern shape, rather than the slender bodies more reminiscent of the baroque guitar. While he is not the originator of fan bracing for the soundboard, his name is associated with it and it is a characteristic feature of his instruments (see Figure 1.2). Many builders still use his fan bracing pattern.

Around the time that Torres introduced his refined guitars, serious composers turned to the instrument in greater numbers. Simultaneously, virtuoso guitarists began to perform publicly, bringing both the music and the instrument itself to a wide audience. Perhaps the final step in this process of the guitar becoming a legitimate classical instrument came when Andres Segovia (1893–1987) began his professional career. He, more than any other classical guitarist, raised the public opinion of the guitar in the early 20th century and firmly established it as a serious instrument.

While the classical guitar was being developed in Europe and adopted by serious musicians, the guitar was also growing in popularity in the United States, though at essentially the opposite end of the social spectrum [6]. Slavery was still practiced in the southern part of the United States in the early and mid 19th century and the slaves taken from Africa brought with them their own musical tradition and instruments. As the slaves partially assimilated into American culture, they also adopted the acoustic guitar.

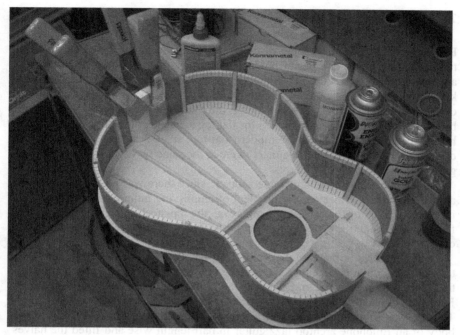

Figure 1.2 Classical Guitar with a Simple Fan Braced Soundboard

It's worth noting at this point that the development of the guitar in the United States proceeded along two lines, Martin and Torres. Christian F. Martin came to the US from Europe in 1833 and by 1838 had established a guitar making business in Nazareth, PA. He had been an apprentice with Johann Georg Stauffer, an accomplished instrument maker in Vienna, and made instruments following Stauffer's designs. These instruments formed the basis for the modern steel stringed acoustic guitar.

Guitars following the Martin pattern were generally the ones adopted by freed slaves after the American Civil War. The styles of music they created developed from the inhumanity of the life forced upon them. Gospel and blues emerged from the American south as a powerful musical force and with them came the acoustic guitar. By the early 20th century, a group of African-American blues musicians, nominally free though still beset by cruel, institutionalized prejudice, established themselves as masters of the acoustic guitar. Musicians such as Robert Johnson, Huddie Ledbetter (also known as Leadbelly) and others brought blues and the precursors of rock and roll to a broad audience.

At that time in the American west, the guitar was also a favorite instrument among cowboys, another group near the bottom of the social strata. The image of the cowboy and his guitar is a familiar one to Americans. Guitars were common in the American west of the late 19th century, at least partly because of the Spanish influence coming through Mexico. The typical cowboy guitar was built along the lines of a Torres classical guitar, though movie and TV cowboys typically played steel stringed (Martin-inspired) instruments.

1.2 The Modern Guitar

By the early 20th century, some of the great American guitar companies had established themselves and began producing instruments that would be familiar to any modern guitarist. Orville Gibson began making guitars and became well-known for large, archtop acoustic guitars [6]. C.F. Martin began what is nearly a dynasty of guitar making. At this writing, the Martin guitar company is being run by C.F. Martin IV and is well known for flat top acoustic guitars like the D-28 that have become icons. Indeed, the modern flat top, steel string acoustic guitar is a descendant of the early Martin guitars in much the same way that modern classical guitars are descended from Torres.

Before the 1920s, all guitars were acoustic since there was no practical way of amplifying them. However, the problem of amplifying guitars was gradually being solved. One of the earliest attempts was the lap-steel guitar (nicknamed 'the frying pan') produced by Rickenbacker in the 1930s. Figure 1.3 shows one of the original patents

One of the most influential innovators was Les Paul [7, 8]. Les Paul (born Lester Polfus) is unique in being a very successful performer, songwriter and inventor. In addition to achieving fame in the United States performing with Mary Ford, he also made one of the early practical electric guitars, the Les Paul 'log'. The log was solid timber fitted with a neck, bridge and pickups. He sawed the body of an Epiphone archtop jazz guitar in half length wise and fitted the halves on either side of the 'log'. The result was a guitar that had no feedback problems and had sustain superior to amplified acoustic guitars. In addition to work with guitars, he also made fundamental advances in recording technology; in 2005, he was inducted into the Inventors Hall of Fame.

It is important to note that magnetic (inductive) pickups like the ones that appeared in the early Gibson and Rickenbacker guitars only work with steel strings. Steel stringed acoustic instruments offered the possibility of being louder than those with gut (or later, nylon) strings, but with the complication of higher string tensions. One of the features introduced by Martin in the early 1900s was the X-braced top capable of resisting increased string tension while still being flexible enough to be an effective sound radiator.

The next big development was the solid body electric guitar. In 1948, Leo Fender introduced the Broadcaster solid body electric guitar. After a trademark dispute, it was renamed the Telecaster in 1951 (Figure 1.4), and became the first commercially successful solid body electric guitar [9]. Because it had a clear tone and was essentially impervious to feedback, it quickly became popular with working musicians.

While electromagnetic pickups had been in use for years, they were generally installed on acoustic instruments. The increase in volume that came with pickups and tube amplifiers caused feedback problems. The body of an acoustic guitar is intentionally flexible so it can produce sound. However, this also makes it respond to the sound field around it – a high enough sound level will make the body vibrate. If there is a pickup on the guitar, this vibration is detected and fed to the amplifier, which then further amplifies the sound and outputs it to a speaker. The speaker then causes the body to vibrate at a higher amplitude and the loop is repeated. The result is the familiar screech of acoustic feedback.

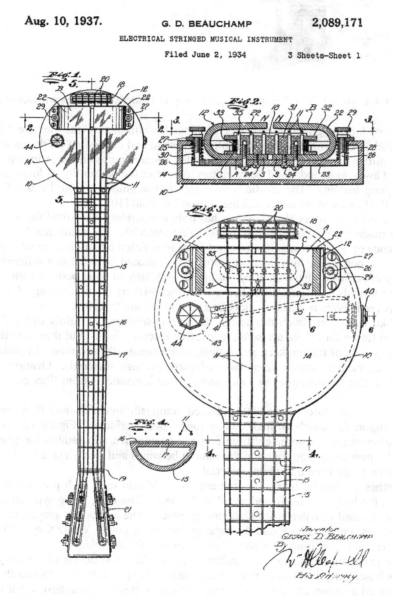

Aug. 10, 1937.   G. D. BEAUCHAMP   2,089,171

ELECTRICAL STRINGED MUSICAL INSTRUMENT

Filed June 2, 1934   3 Sheets—Sheet 1

Figure 1.3 Drawing From Patent Application for the Rickenbacker Electric Guitar (Wikipedia Commons, image is in the public domain)

Figure 1.4 Re-issue 1952 Telecaster (image courtesy of Fender Guitars, www.fender.com)

The solid body guitar solved this problem by eliminating the flexible body completely, replacing it with a rigid block of wood and depending on the pickups and amplifier to make all the sound. Because kinetic energy in the strings is not radiated away as sound, solid body guitars also offer the possibility of long decay times (long sustain). The classic solid body electric guitars are the Fender Telecaster, the Fender Stratocaster and the Gibson Les Paul [10].

Since the body of an electric guitar has only a secondary acoustical function, it can be made in any shape and of any suitable material. The result has been the emergence of the electric guitar as an art form in which the body is as much a palette as it is part of the instrument (Figure 1.5). It should be noted that differences in body materials do have an effect on the tonal quality of solid body electric guitars, but it tends to be subtle. Still, accomplished players often have very definite preferences in body materials – alder and swamp ash are favorites.

Unlike violins in which the design of most instruments is rigidly dictated by tradition (to be fair, there are some innovative electric violins, but they are still a small minority of the market), guitar designs are constantly evolving. In particular, designers and manufacturers have adopted synthetic materials. Ovation was the first major manufacturer to make widely-used instruments from fiber composites.

Other manufacturers have since adopted composite materials and their use is increasing as the supply of high quality tone woods dwindles. Figure 1.6 shows an instrument made completely from graphite composites. The author has played a similar instrument and found to the tone to be warm and rich – comparable to a fine guitar made from traditional materials.

Another interesting line of instruments from Martin uses high pressure laminates in the body and laminated wood in the neck. The tops, being synthetic, are sometimes used as palettes for elaborate graphics. The author has played several of these instruments and found the tonal quality to be good. While they are relatively new at this writing, they should prove to be quite durable.

One interesting experiment in guitar design is the construction of a family of guitars based on principles of structural dynamics [11]. Graham Caldersmith has constructed a group of four guitars consisting of a treble, a standard, a baritone and a bass. The treble has string frequencies 50% higher than those of a standard guitar (this is usually called a musical fifth, see Chapter 2). The baritone has strings tuned 33% lower than a standard guitar (decreased by a musical fifth) and the bass has strings tuned 50% lower (decreased by an octave). This effort is analogous to a project in which a group of eight instruments in the violin family were designed and built using dynamic scaling concepts [12].

Figure 1.5 Bo Diddley Playing His Signature Rectangular Guitar (Wikipedia Commons, image is in the public domain)

Figure 1.6 An Acoustic Guitar Made Entirely of Graphite Composites (image courtesy of RainSong Guitars, www.rainsong.com)

A recent development in guitar design is the acoustic-electric hybrid instrument. These can take many forms, but are generally instruments with sophisticated pickups, pre-amplifiers and sometimes graphic equalizers mounted in a thin, semi-acoustic body. The body is hollow and the soundboard is flexible so that the instrument radiates sound even when not amplified. The bodies are thin for comfort and the resulting un-amplified sound can be thin and unsatisfying. These instruments are intended to be amplified and, when plugged in, can sound very good indeed. They have the additional advantage that they can be produce both acoustic and electric sounds which allows a stage musician to use the same instrument for different pieces.

Figure 1.7 Two Thin Body Acoustic-Electric Guitars (photo by the author)

2 Acoustics and Musical Theory

Pythagoras appears to have been the first to notice that the pitch of strings varies according to their length and that certain combinations of string lengths produced notes that sounded good when combined [13]. He divided a vibrating string into two parts so that each part could vibrate and produce its own frequency. He also found that, if the lengths of the two parts were related by simple ratios like 2:1 and 3:2, the resulting combination was pleasing. Modern terminology labels these combinations consonant.

Some of the most familiar features of the modern guitar are dictated by the structure of music and by the fact that the frequencies of strings vary according to length. The foundations of western music were in place certainly by the end of the Renaissance [14], so the ideas and the terminology are very old. They have an ancient and solid feel about them that contrasts with the cool, technical descriptions of the underlying concepts.

Music, at its most fundamental level, is just a progression of carefully structured sounds. So, along with an understanding of the music should also come an understanding of some basic ideas in acoustics. In particular, it is helpful to see how amplitudes are quantified and how the human ear responds to sounds.

2.1 Basics of Music

All music, no matter what kind, is made of the same building blocks – individual notes. A pure tone of any frequency in the human hearing range can be made with a signal generator and a speaker. Idealized notes in a musical scale are sounds consisting of pure tones (sine waves) combined with harmonic overtones. They have specific frequencies chosen very precisely from the continuum of possible frequencies in the human hearing range. We are all different, but, based on extensive measurements, the human hearing range is generally assumed to be 20 Hz – 20 kHz [15].

From the beginnings of music, people realized that certain combinations of notes sounded good together. These combinations became widely used and were given names to make them easy to work with. In the early 1600s Galileo built on the work of Pythagoras and showed that the length of a string under a constant tension is inversely proportional to its lowest natural frequency [16]. From this discovery came the realization that these pleasing combinations of notes corresponded to frequencies that were the ratios of small integers like 2:1 and 3:2.

In western music, the notes are related to each other by given ratios [17]. Thus, if we know the absolute frequency of a single note and the frequency ratios that define the relationships between that note and the others, we can find the frequency of every possible note. By international agreement, the frequency standard used for music is the A note, now defined to be 440 Hz [18].

R.M. French, *Engineering the Guitar*,
DOI: 10.1007/978-0-387-74369-1_2, © Springer Science+Business Media, LLC 2009

The largest interval is an octave and represents a doubling of frequency. The A note with a frequency of 440 Hz is more precisely called A_4. The note one octave higher in pitch is called A_5 and has a frequency of 880 Hz. Similarly, the note one octave lower is called A_3 and has a frequency of 220 Hz.

The octave is, in turn made of smaller intervals. In western music, the smallest possible interval between two notes is called variously a semi-tone, a half tone or a half step. In this book, this interval will be called a half step. In western music, there are 12 half steps in an octave, but other cultures have used other numbers; there is no physical requirement to use 12 half steps. Predictably, an interval of two half steps is called a whole step or a whole tone. For consistency, here it will be called a whole step.

The various intervals between two notes have unique names to identify them. For each interval, there is a corresponding inversion and the sum of any interval and its inversion is an interval of one octave (12 half steps). Figure 2.1 shows the intervals and their names.

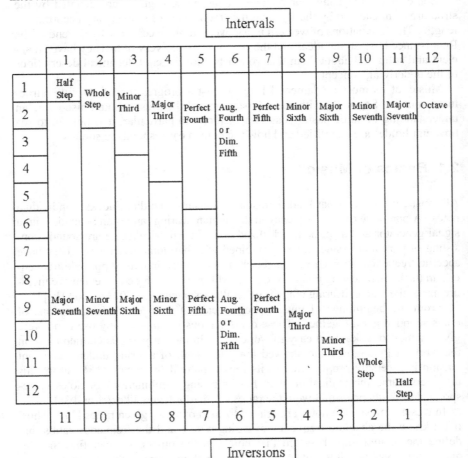

Figure 2.1 Names of musical intervals

While, there are 12 possible notes in an octave, the very name octave suggests that only eight of them are used. The scale is the next level of musical structure and dictates which eight of the 12 notes are used. A musical scale is just a fixed pattern of intervals [19]. There are several different scales, though even non-musicians might be familiar with the names of the most widely used, the major scale and the minor scale. The structure of the major and minor scales are shown in Figure 2.2

In order to conveniently describe the notes and the intervals between them, they had to be given names. The names of the notes themselves are simply the letters A through G with the names repeating every octave.

Major Scale

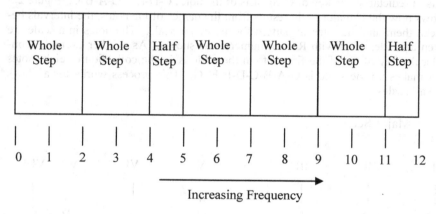

Minor Scale

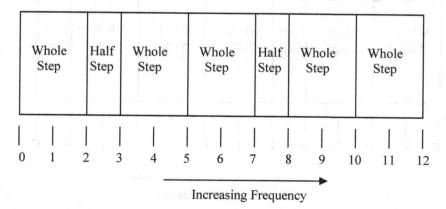

Figure 2.2 The patterns of whole steps and half steps that form major and minor scales

In order to construct a specific scale from the patterns shown above, we need to know how the seven unique letter names are distributed among the 12 half step interval that forms an octave. There is a whole step between all the letter names except for B-C and E-F, which are separated by half steps. Additionally, notes can be modified by a sharp,♯, which raises the note by a half step or by a flat,♭ , which lowers the note by a half step.

A scale can start on any note and it helps to give the scale a name that compactly indicates which notes are in it. This name is called the key and it is simply the first note in the scale. For example, the name C major tells the musician everything needed about which notes will appear in a piece of music written in this key.

Beginning musicians often play in the key of C major since it has no sharps or flats. Predictably, the key of C consists of the notes C-D-E-F-G-A-B-C. Figure 2.3 shows how the C major scale results from the names of the notes, the intervals between them and the interval pattern for the major scale. The notes in a scale are often also identified with Roman numerals as shown. As another example, consider the key of G. If the first note in the scale is G, the collection of eight notes that makes up the scale is G-A-B-C-D-E-F♯-G. This process works for all keys and all scales.

Major Scale

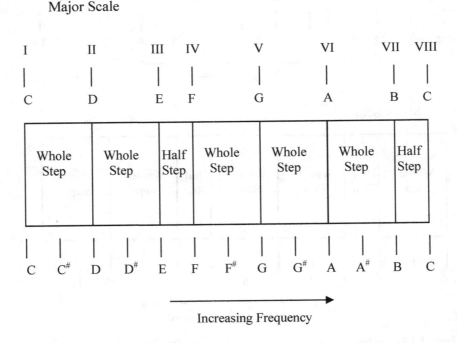

Increasing Frequency

Figure 2.3 The notes in the C major scale

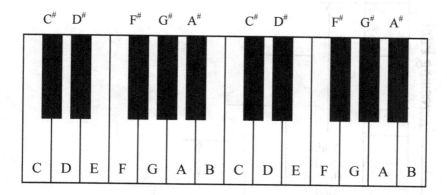

Figure 2.4 Piano Keyboard with Names of Notes

The notes and the intervals between them are laid out in an intuitive way on the piano keyboard. Figure 2.4 shows a section of a piano keyboard and the corresponding notes. It is clear from the way the keyboard is arranged that playing in the key of C major means only playing the white keys.

For simplicity, the discussion here is confined to major scales. However, other scales can be defined using the same kind of pattern used to define the major scale. A scale made from a group of seven notes consisting of five whole steps and two half steps, in which the half steps are separated by either two or three whole steps is called a Diatonic scale [20, 21].

Diatonic scales are the foundation of Western music and the piano keyboard is laid out in Diatonic intervals. Figure 2.5 shows the seven diatonic scales that can be made from a repeating pattern of half and whole steps. By far, the most widely used are the major and the natural minor. However, guitarists, particularly those who play lead guitar in a band, are often students of the different musical scales. It is perhaps helpful to see that they are all logically related to one another; indeed, they are formed simply by choosing different starting points on the repeating pattern of intervals used to define the major scale.

There is one more piece of the scale puzzle – the numerical relationship between the notes. Music has evolved over thousands of years, and, for historical reasons there is more than one way to define the frequency ratios between the notes [22, 23]. While this might seem to be an arcane diversion, it is central to the development of the guitar.

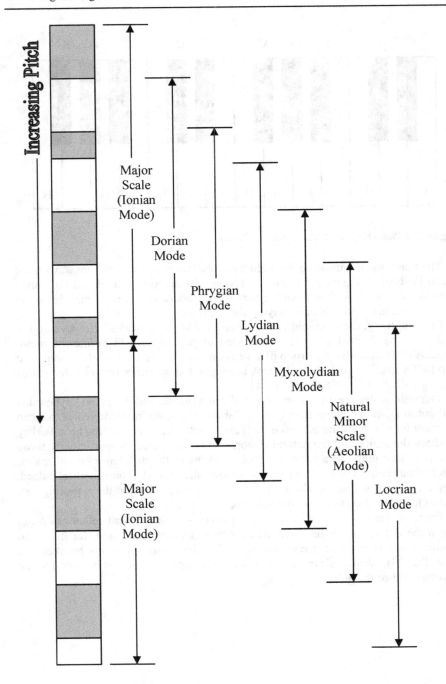

Figure 2.5 Structure of Diatonic Scales

2.2 Scales and Temperament

No matter which scale is being used, there needs to be a precise description of the frequency ratios. The most consonant ratio of frequencies is the octave, a 2:1 frequency ratio. The next two are 3:2, a perfect fifth, and 4:3, a perfect fourth. Note from Figure 2.1 that the combination of a perfect fifth (7 half steps) and a perfect fourth (5 half steps) is an octave (12 half steps). Thus, going up a fourth results in the same letter as going down a fifth [17]. This is true no matter what starting point is chosen. Furthermore, any note can be reached from any other note by moving up in steps of fifths only or by moving down in steps of fourths only. Figure 2.6 shows the first two steps taken when increasing pitch in steps of a fifth.

A Pythagorean scale results when frequency ratios based on increasing fifths (or decreasing fourths) are used to determine the notes. For example, the interval between C and D is two half tones. D can also be reached by increasing pitch by two successive fifths. Thus the frequency of the C note is raised by $3/2 \times 3/2 = (3/2)^2 = 9/4$. However, since 9/4 is more than 2, the resulting D is an octave high.

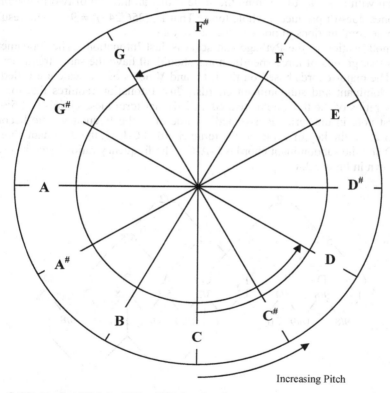

Figure 2.6 Increasing Pitch by Fifths (3:2) Starting from C

I	II	III	IV	V	VI	VII	
C	D	E	F	G	A	B	C
1	9/8	81/64	4/3	3/2	27/16	243/128	2
	9/8	9/8	256/243	9/8	9/8	9/8	256/243

Figure 2.7 Frequencies and Intervals on the Pythagorean Scale (after Rossing, Moore and Wheeler [17])

Reducing it by an octave is equivalent to dividing the frequency by 2, so the frequency for D is 9/8 that of C. The remainder of the scale can be developed in this manner so that the frequency ratios for the Pythagorean scale are as shown in Figure 2.7.

There are only two different intervals in the Pythagorean scale shown in Figure 2.7: 9/8 and 256/243. These are the whole step and the half step respectively. The problem with this way of defining the scale is that an interval of two Pythagorean half tones doesn't produce a whole tone. That is, $(256/243)^2 \neq 9/8$. The result is that some combinations of notes sound out of tune.

A modification to the Pythagorean scale is Just Intonation. The fundamental idea is that groups of three specific notes should all have the same frequency ratios. The major chords based on the I, IV and V notes of the scale are called the tonic, dominant and subdominant chords. Just Intonation requires that the frequency ratios of the three notes (called a triad) that form these chords is 4:5:6. If the first note in the series is assigned a value of 1, the frequency ratio becomes 1:5/4:3/2. In the key of C major, the tonic chord is C-E-G, the dominant chord is G-B-D and the subdominant chord is F-A-C. The frequency ratios in the just scale are shown in Figure 2.8.

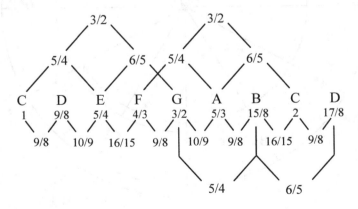

Figure 2.8 Frequency Ratios in the Just Scale (C Major)

While this system has an aesthetic simplicity, it presents the musician with some serious practical problems. For example, not all intervals of 7 half steps are perfect fifths with a frequency ratio of 3:2. The frequency ratio of D:A in the key of C shown in Figure 2.8 is

$$\frac{10}{9}\times\frac{16}{15}\times\frac{9}{8}\times\frac{10}{9}=\frac{40}{27}=1.4815 \qquad (2.1)$$

While the resulting frequency ratio is close to 3:2, the difference is clearly audible. This effect is even more of a problem when one considers that not all instruments in an orchestra play in the same key. Consider the Just Scale in $B^{\flat}$ as shown in Figure 2.9.

In the case of the key of $B^{\flat}$, the frequency ratio of D:A really is 3:2.

$$\frac{16}{15}\times\frac{9}{8}\times\frac{10}{9}\times\frac{9}{8}=\frac{3}{2}=1.500 \qquad (2.2)$$

As a result, it is not practical for an orchestra to be composed of instruments with just intonation; the resulting sound would be unpleasantly dissonant.

The practical solution is to slightly modify or temper the frequency ratios of the just scale. There are several common temperament schemes and the one used for guitars, called equal temperament, starts with a fixed half step frequency ratio – one that is identical for all half step intervals, no matter what the root note is. All the other intervals are then derived from this grounding assumption. The equal tempered scale is not perfect either, and can produce some unpleasantly dissonant combinations of notes. In practice, different temperaments are used as necessary.

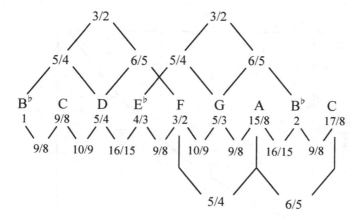

Figure 2.9 Frequency Ratios in the Just Scale ($B^{\flat}$ Major)

An equal tempered scale uses a fixed proportion to define the pitch change between notes. If some note has a frequency, f_0, and the proportional change in pitch is r, the frequency of successive notes is

$$f_n = r^n f_0 \qquad (2.3)$$

Thus, the frequencies as the pitch increases by half steps are f_0, rf_0, $r^2 f_0$ and so on. The twelfth note in the series is a full octave above the original one. Since increasing pitch by an octave doubles the frequency, it is easy to find r.

$$f_{12} = 2f_0 = r^{12} f_0$$

$$2 = r^{12} \qquad (2.4)$$

$$r = \sqrt[12]{2} \cong 1.0594631$$

The frequency relationship can also be written in term of logarithms

$$f_n = r^n f_0$$

$$\frac{f_n}{f_0} = r^n$$

$$\log\left(\frac{f_n}{f_0}\right) = \log\left(r^n\right) \qquad (2.5)$$

$$\log f_n - \log f_0 = n \log r$$

Figure 2.10 shows a plot of the frequency ratios over two octaves on linear axes.

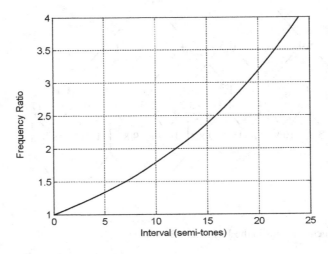

Figure 2.10 Frequency Ratios for Half Step Intervals

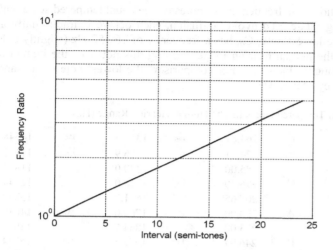

Figure 2.11 Frequency Ratios for Half Step Intervals – Log Scale

Since the function defining frequency ratios is a power series, the curve is straight when plotted on a semi-log scale as shown in Figure 2.11

To make the concept of frequency ratios more concrete, Table 2.1 shows a complete octave using middle A (440 Hz) as a starting point

Table 2.1 Frequencies and Ratios for an Octave

Notes in Major Scale	Interval (Half Steps / Note)	Interval Name	Equal Tempered Frequency Ratio	Equal Tempered Frequency (Hz)	Just Frequency Ratio	Just Frequency (Hz)
1	0 / A		1.0000	440.00	1	440.00
	1 / A#	Half Step	1.0595	466.16		
2	2 / B	Whole Step	1.1225	493.88	9/8	495.00
	3 / C	Minor Third	1.1892	523.25	6:5	528.00
3	4 / C#	Major Third	1.2599	554.37	5/4	550.00
4	5 / D	Perfect Fourth	1.3348	587.33	4/3	586.67
	6 / D#	Augmented Fourth/ Diminished Fifth	1.4142	622.25	64/45	625.78
5	7 / E	Perfect Fifth	1.4983	659.26	3/2	660.00
	8 / F	Minor Sixth	1.5874	698.46	8:5	704.00
6	9 / F#	Major Sixth	1.6818	739.99	5/3	733.33
	10 / G	Minor Seventh	1.7818	783.99	16/9	782.22
7	11 / G#	Major Seventh	1.8877	830.61	15/8	825.00
8	12 / A	Octave	2.0000	880.00	2	880.00

Table 2.2 shows the frequencies of notes on the equal tempered scale within the human hearing range. To avoid confusion, each octave is traditionally given a subscript. The lowest note, C_0, has a frequency of 16.352 Hz (slightly below the beginning of the human hearing range). Using this notation, the human hearing range spans roughly $E_0 - E_{10}$. For comparison, the lowest note on a Piano is A_0 with a frequency of 27.5 Hz.

Table 2.2 Equal Tempered Notes In The Human Hearing Range (Hz)

E_0	20.602	E_3	164.81	E_6	1318.5	E_9	10548
F_0	21.827	F_3	174.61	F_6	1396.9	F_9	11175
	23.125		185.00		1480.0		11840
G_0	24.500	G_3	196.00	G_6	1568.0	G_9	12544
	25.957		207.65		1661.2		13290
A_0	27.500	A_3	220.00	A_6	1760.0	A_9	14080
	29.135		233.08		1864.7		14917
B_0	30.868	B_3	246.94	B_6	1975.5	B_9	15804
C_1	32.703	C_4	261.63	C_7	2093.0	C_{10}	16744
	34.648		277.18		2217.5		17740
D_1	36.708	D_4	293.66	D_7	2349.3	D_{10}	18795
	38.891		311.13		2489.0		19912
E_1	41.203	E_4	329.63	E_7	2637.0	E_{10}	21096
F_1	43.654	F_4	349.23	F_7	2793.8		
	46.249		369.99		2960.0		
G_1	48.999	G_4	392.00	G_7	3136.0		
	51.913		415.30		3322.4		
A_1	55.000	A_4	440.00	A_7	3520.0		
	58.270		466.16		3729.3		
B_1	61.735	B_4	493.88	B_7	3951.1		
C_2	65.406	C_5	523.25	C_8	4186.0		
	69.296		554.37		4434.9		
D_2	73.416	D_5	587.33	D_8	4698.6		
	77.782		622.25		4978.0		
E_2	82.407	E_5	659.26	E_8	5274.0		
F_2	87.307	F_5	698.46	F_8	5587.7		
	92.499		739.99		5919.9		
G_2	97.999	G_5	783.99	G_8	6271.9		
	103.83		830.61		6644.9		
A_2	110.00	A_5	880.00	A_8	7040.0		
	116.54		932.33		7458.6		
B_2	123.47	B_5	987.77	B_8	7902.1		
C_3	130.81	C_6	1046.5	C_9	8372.0		
	138.59		1108.7		8869.8		
D_3	146.83	D_6	1174.7	D_9	9397.3		
	155.56		1244.5		9956.0		

The frequencies calculated using the equal tempered scale are slightly different – less than 1% – from those using a just scale. The guitar player is basically trading the ability to play a just scale for the ability to use a regularly spaced fret pattern and the ability to play easily up and down the neck. Another solution to the problem is to dispense with frets as on instruments of the violin family. Thus, a skilled violinist can have some of the same flexibility of the guitar without being bound to equal temperament. This too comes at a price, though. With no frets, the accuracy of any note is completely dependent on the skill of the player. Any parent of a young strings (violin, viola, cello or bass) player can tell you that the early stages are as much a chore for anyone within earshot as for the child.

In addition to standardized frequencies for the individual notes, the six strings of the guitar also have a standard tuning pattern. The standard tuning pattern is EADGBE and the corresponding frequencies are shaded in Table 2.2. A large number of alternate tunings exist with 'Drop D' (DADGBE) and DADGAD being among the more popular. However, a large majority of guitar music is written for standard tuning.

Guitars are used to play both single notes and chords. In fact, rhythm guitarists are mostly known as chord players. A chord is just a collection of at least three notes from a diatonic scale which are played together (three note groups are called triads). Each different kind of chord uses a different group of notes. For instance, major chords use notes 1, 3 and 5, minor chords use notes 1, ♭3 and 5 from the major scale or notes 1, 3 and 5 from the minor scale. Thus, a C major chord is formed from the notes labeled I, III and V from the C major scale shown in Figure 2.3 – the notes C, E and G. Consider a major chord, made using the 1, 3 and 5 notes in a major scale using A as the starting point; the notes in the triad are A, $C^{\#}$, E.

From a strictly mechanical point of view, a guitar is a device that connects strings under tension to a resonator with flexible walls and offers a convenient way to shorten the strings to raise their frequencies. The frets on a guitar neck are there to stop the strings, effectively shortening them. They are arranged so that the space between two successive frets on the same string is always a half step. To figure out the fret spacing, we need to know the relationship between the frequency (in radians/sec) of a vibrating string and its length

$$\omega_n = \frac{n\pi}{L}\sqrt{\frac{T}{\rho}} \tag{2.6}$$

where T is the tension in the string and ρ is the mass density of the string [17]. The first frequency, ω_1, is the fundamental frequency and the frequencies ω_2, ω_3 and so on are called overtones or higher harmonics.

The frequency with which a string vibrates is inversely proportional to the length of the string as long as nothing else changes. For instance, doubling the length of a string would lower the frequency by half or one octave. Knowing this, it is not hard to write down the expression for the position of the frets on the neck. If L is the scale length (distance from nut to bridge), the distance, d, from the bridge to the nth fret is

$$d_n = \frac{L}{r^n} \qquad (2.7)$$

The distance from the nut to the nth fret is

$$d_n = L - \frac{L}{r^n} = L\left(1 - \frac{1}{r^n}\right) \qquad (2.8)$$

Since half step frequency ratios defined using just intonation are slightly different depending on the root note, it would be impossible to have a regular fret spacing on the guitar neck. An equal tempered scale allows regular fret spacing, though the resulting notes are not at the exact frequency as they would be with just intonation. Having all the frets spaced the same way makes the guitar very versatile since there are often several ways of playing the same note on different strings. This allows interesting combinations of sounds that can give a skilled player added expressive freedom. Conversely, on a piano, there is one key for every note. The combinations of notes that can be played are limited by what an individual player can reach. Indeed, some piano pieces are intended for two people playing at once on the same piano.

There are many different scale lengths in use. However, two common ones are 25.5 in (648 mm associated with Fender guitars) and 24.75 in (629 mm, associated with Gibson guitars). The fret spacings for those two scale lengths are presented in Table 2.3. Fret spacings for other scale lengths can be calculated readily using a spreadsheet program.

The spacings presented here use the factor $r = 2^{1/12}$, but luthiers sometimes use a simplification called 'The Rule of 18.' 18/17 is approximately 1.0588 and is close enough to $2^{1/12}$ that is can be used to lay out frets. The string length is simply reduced by 1/18 of its length to increase the frequency by an equal-tempered half step.

Table 2.3 Fret Spacing for Two Common Scale Lengths (inches)

Fret	Fender (25.5 in) From Bridge	Fender (25.5 in) From Nut	Gibson (24.75 in) From Bridge	Gibson (24.75 in) From Nut
0 (nut)	25.500	0.000	24.750	0.000
1	24.069	1.431	23.361	1.389
2	22.718	2.782	22.050	2.700
3	21.443	4.057	20.812	3.938
4	20.239	5.261	19.644	5.106
5	19.103	6.397	18.541	6.208
6	18.031	7.469	17.501	7.249
7	17.019	8.481	16.519	8.231
8	16.064	9.436	15.592	9.158
9	15.162	10.338	14.716	10.034
10	14.311	11.189	13.890	10.860
11	13.508	11.992	13.111	11.639
12	12.750	12.750	12.375	12.375
13	12.034	13.466	11.680	13.070
14	11.359	14.141	11.025	13.725
15	10.721	14.779	10.406	14.344
16	10.120	15.380	9.822	14.928
17	9.552	15.948	9.271	15.479
18	9.016	16.484	8.750	16.000
19	8.510	16.990	8.259	16.491
20	8.032	17.468	7.796	16.954
21	7.581	17.919	7.358	17.392
22	7.156	18.344	6.945	17.804
23	6.754	18.746	6.555	18.195
24	6.375	19.125	6.188	18.563

Table 2.4 shows the frequencies on a guitar neck. The lowest note on the neck is E on the 6th string – 82.4 Hz. Doubling that frequency gives 164.8 Hz. Doubling again gives 329.6 Hz, so there are two octaves between string 1 and string 6. If your guitar has 24 frets, you could raise string 1 by another two octaves. Thus, the instrument has a 4 octave range.

Table 2.4 Frequencies On A Guitar Neck (Frequencies in Hz)

Fret	String 6	String 5	String 4	String 3	String 2	String 1
nut	82.400	110.00	146.80	196.00	246.90	329.60
1	87.300	116.54	155.53	207.65	261.58	349.20
2	92.491	123.47	164.78	220.00	277.14	369.96
3	97.991	130.81	174.58	233.08	293.61	391.93
4	103.82	138.59	184.96	246.94	311.07	415.27
5	109.99	146.83	195.95	261.63	329.57	439.96
6	116.53	155.56	207.61	277.19	349.17	466.12
7	123.46	164.81	219.95	293.67	369.93	493.84
8	130.80	174.61	233.03	311.13	391.93	523.21
9	138.58	185.00	246.89	329.63	415.23	554.32
10	146.82	196.00	261.57	349.23	439.93	587.28
11	155.55	207.65	277.12	370.00	466.09	622.20
12	164.80	220.00	293.60	392.00	493.80	659.20
13	174.60	233.08	311.06	415.31	523.16	698.40
14	184.98	246.94	329.55	440.00	554.27	739.93
15	195.98	261.63	349.15	466.17	587.23	783.93
16	207.63	277.18	369.91	493.89	622.15	830.54
17	219.98	293.66	391.91	523.26	659.14	879.93
18	233.06	311.13	415.21	554.37	698.34	932.25
19	246.92	329.63	439.90	587.34	739.86	987.68
20	261.60	349.23	466.06	622.26	783.86	1046.4
21	277.16	369.99	493.77	659.26	830.47	1108.6
22	293.64	392.00	523.14	698.46	879.85	1174.6
23	311.10	415.30	554.24	740.00	932.17	1244.4
24	329.60	440.00	587.20	784.00	987.60	1318.4

Finally, it is worth noting that the guitar is a C instrument. This distinction is often mentioned in books on musical instruments, but not always explained. A C instrument is one for which playing a note called C yields the standard frequency, also called also called concert pitch, for the C note as shown in Table 2.2. This would seem to be trivial except that not all instruments are C instruments. These are called transposing instruments; the guitar is a non-transposing instrument and guitar music can be written in concert pitch.

Most clarinets, for example, are $B^\flat$ instruments (though there is such a thing as a C clarinet). Thus, playing the note called C on a standard clarinet would actually give a frequency corresponding to $B^\flat$, two half steps lower ($C \rightarrow B \rightarrow B^\flat$) than C. A piece written for clarinet and guitar would then have the two instruments playing in different keys so that they would be in tune with one another.

2.3 Quantifying Sound

Like most other physical phenomena, sound has specific ways of being quantified. Sound is the name we give to time-varying air pressure within a frequency range of 20 Hz–20 kHz. However, knowing the frequency alone isn't enough; one must also know the amplitude (volume) of the sound pressure. The SI unit for pressure is the Pascal (Pa), which is 1 N/m^2, but amplitude is usually reported in decibels (dB).

The dB is different from most other familiar units because it is dimensionless and it is logarithmic. The definition of a dB is

$$X_{dB} = 20\left(\log_{10} \frac{x_{RMS}}{x_{ref}} \right) \tag{2.9}$$

Where X_{RMS} is the RMS average of the signal and X_{ref} is a reference value with the same units as the signal. The RMS average is required for sound measurements because the mean of any sinusoidal signal that oscillates about zero is zero no matter what its amplitude might be. RMS captures the amplitude of a signal that oscillates about zero. By international agreement [24], the standard reference value for sound measurements is 20 μPa (20 × 10^{-6} Pa). The RMS pressure in Pa is called sound pressure. The level reported in dB is called sound pressure level (SPL). For example, an RMS sound pressure of 1 Pa gives a SPL of 93.98 dB. It is useful to develop some intuition about what a sound pressure level means in real life. Table 2.5 shows some approximate sound levels for common activities.

Table 2.5 Environmental Sound Pressure Levels

Sound	SPL (dB)
Permanent deafness after short exposure	150
Custom car with 1,000 watt stereo, interior	140
Pain Threshold	130
Jet takeoff at 100m	120
OSHA max allowable, any duration	115
Front rows of rock concert	110
Outdoor rock concert	105
Large Orchestra	98
OSHA max allowable, 8 hours	90
OSHA 'action level' for 8 hour exposure	85
Car interior at highway speed	70
Conversational speech	60
Private office	50
Bedroom at night	30
Whisper	20
Threshold of hearing for a child	0

2.4 Sound Radiation

Radiation is the process of creating sound waves and letting them propagate away from the source. A guitar has two types of radiation sources, the soundhole and the vibrating plates. The top plate is generally assumed to radiate much more sound energy than the back.

Acousticians often speak in terms of how many poles a radiator has. A monopole is a point source that radiates out in all directions. A dipole is formed by two closely-space monopoles and so on [25, 26]. Figure 2.12 shows the wave pattern from an ideal monopole radiator in two dimensions. The center of the radial pattern is the location of the source and the waves radiating out represent positive and negative air pressure at an instant in time.

Figure 2.13 shows the analogous pattern from an ideal dipole whose two sources are 180° out of phase. Again, this is a two-dimensional analogy since it is quite difficult to illustrate the actual three-dimensional pressure field on a sheet of paper. Note the line on which the amplitude of the wave is zero. An observer positioned exactly on that line wouldn't be able to detect the wave. Thus, if this were an acoustic wave, a listener standing on the zero amplitude line wouldn't be able to hear the signal (assuming no reflections). These two patterns relate directly to the guitar since the different parts of the top can vibrate out of phase with one another.

The soundhole acts roughly like a monopole radiator for the first coupled airbody mode of a guitar [27]. Mathematically, it can be treated as a piston in the same manner as a speaker cone in a cabinet. In the region near a piston (the near field), the sound is concentrated on the axis of motion. Far from the piston (the far

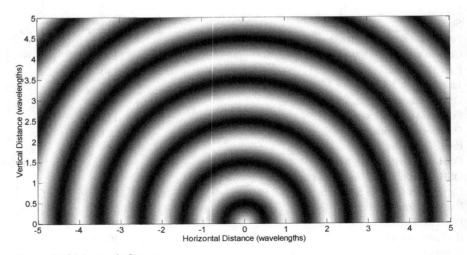

Figure 2.12 Monopole Source

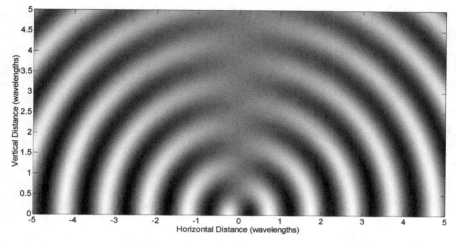

Figure 2.13 Dipole Source

field), the sound is approximately omnidirectional and is only a weak function of the angle from the piston axis [28]. For a guitar, the far field might be defined as the region more than a few body lengths from the top. While a listener could not conveniently be within the near field, it is certainly possible to place a microphone there.

Acoustic radiation from the body is strongly conditioned by the mode shapes of the top. In a manner analogous to the example shown in Figure 2.13, acoustic radiation from adjacent areas of the top that move out of phase with one another may tend to cancel. The shape of a guitar top is difficult to describe mathematically; for simplicity, we'll consider rectangular plates with about the same aspect ratio as a guitar top. The math becomes much easier without changing the underlying concept.

The mode shapes of a simply supported plate (one whose edges can rotate, but not move up or down) are a product of sine waves. The modal velocity distribution for a plate with side lengths a and b is

$$v(x, y) = V \sin(p\pi x / a)\sin(q\pi y / b) \qquad \begin{cases} 0 \le x \le a \\ 0 \le y \le b \end{cases} \qquad (2.10)$$

Where V is the amplitude and p and q define the number of anti-nodes in the x and y direction. A node line is a collection of points that don't move up or down and an anti-node is a point at which motion is a maximum. For the first mode, $p=1$ and $q=1$. This is called the 1,1 mode [29] as shown in Figure 2.14. The 1,1 mode has no interior node line. The 1,2 mode has one interior node line and two anti-nodes and so on. Note that there is more than one method of naming plate modes in the literature. Sometimes, they are named according to the number of internal node lines. Thus, the mode labeled here as (1,1) would be labeled (0,0). The underlying mechanical principles are unchanged by the naming convention.

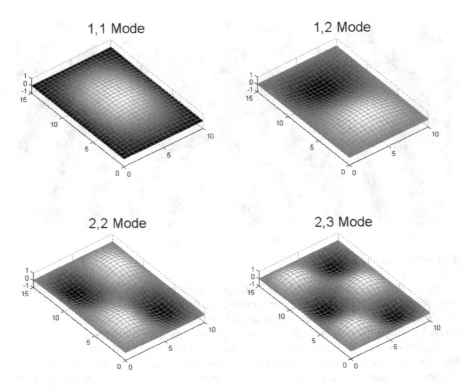

Figure 2.14 Rectangular Plate Modes

The moving portions of the rectangular plate act in the manner of pistons to create pressure waves (sound) that can then radiate out into the space in front of and behind the plate. The 1,1 mode acts as a monopole radiator since the entire plate moves in phase and there are no internal node lines (lines where there is no motion). The 1,2 mode acts as a dipole radiator. However, since the upper and lower portions of the plate are 180° out of phase, the radiated sound field is more complicated. Because of the phase difference, the transient pressures generated at the surface of the upper and lower parts of plate are of opposite signs. This basic idea extends to the higher modes in which there are more and smaller areas of the plate out of phase with each other.

The definitions of the near field and far field are a little loose, since there is no step change in the sound field as one moves farther from the instrument. If we assume the moving portion of a guitar soundboard is 200mm × 250 mm, the near field might be the region closer than a few hundred mm. In the near field, the contributions of the different parts of the vibrating plate are clear.

The region that is significantly farther from the source than the source size is called the far field. In the far field, the dimensions of the source are small enough to be ignored, so calculations and measurements become easier. For a 200 mm × 250 mm

Figure 2.15 Approximate Radiating Areas of a Small Guitar

plate, the far field is the region significantly more than a few hundred mm from the plate. In practice, the far field might be a meter or more from the plate.

Because of the internal structure of a guitar, only the portion of the soundboard in the lower bout generally moves enough to be an effective radiator at low frequencies (Figure 2.15)

2.5 Human Perception of Sound

A microphone is just a sensor that turns sound pressure into a proportional, time varying voltage. The ideal microphone senses all frequencies between 20 Hz and 20 kHz with equal gain so it can perfectly reproduce the part of the sound field it hears. High quality microphones used for acoustic testing come close to this ideal.

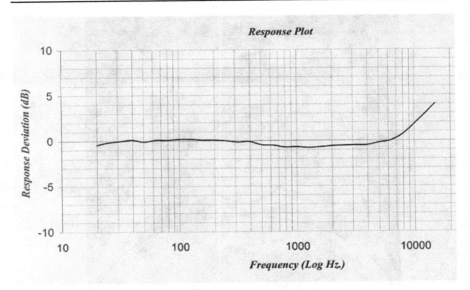

Figure 2.16 Calibration Chart for a Microphone

Figure 2.16 shows the calibration chart for a piezoelectric microphone. The response curve is nearly flat (± 1 dB) from 20 Hz to 8 kHz, but deviates significantly above 10 kHz. Other types of microphones have flat response throughout the human hearing range.

The human ear, though, doesn't have this relatively flat frequency response. The outer ear (called the pinna, this is the part you can see) and the inner ear have their own frequency response characteristics and the result is that people are not equally sensitive to all frequencies [30]. While everyone is different, people typically have a peak in their acoustic sensitivity near 1000 Hz. Additionally, human sensitivity to sound generally decreases significantly below 200 Hz and above 10 kHz. Of course, hearing damage from exposure to loud noises and normal biological variations can change this.

There have been many attempts to correct SPL to match human perceptions of sound levels as functions of both frequency and amplitude. The simplest, and probably most widely used, is the A-weight filter [31]. This is a simple weighting function that attenuates or amplifies a signal based only on its frequency. When sound pressure level is calculated on a signal that has been modified with an A-weight filter, the result is expressed in dB(A). Figure 2.17 shows the frequency response of the A-weight filter.

A change of 0 dB corresponds to multiplying the signal by a factor of 1, so it's clear that the weighting is 1.00 at 1000 Hz. This is why many microphone calibrators generate a 1000 Hz signal. Also, from the curve plotted on linear axes, one can see that the A-weight filter can be crudely approximated by a 1000 Hz low pass filter, particularly if the signal doesn't have content at higher frequencies. Note also how insensitive human ears are to very low frequencies. At 40 Hz, the amplitude is reduced by about 98%.

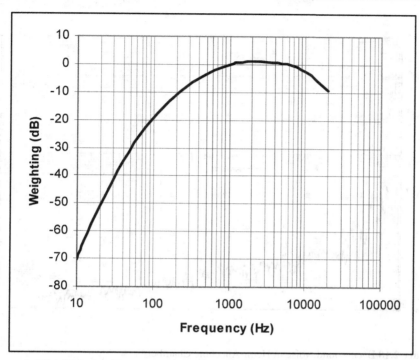

Figure 2.17(a) Frequency Response of the A-Weighting Filter, Log Frequency

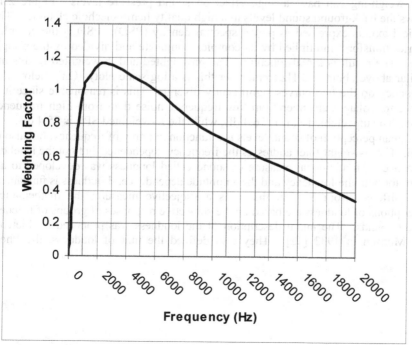

Figure 2.17(b) Frequency Response of the A-Weighting Filter, Linear Frequency

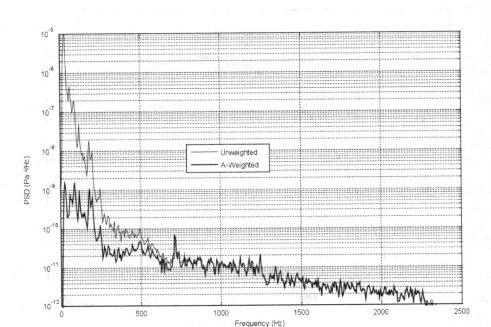

Figure 2.18 Background Noise in Hemi-Anechoic Chamber

A-weighting can have a huge effect on sound pressure levels. Figure 2.18 shows the background sound levels in a high quality hemi-anechoic chamber. The vertical axis is expressed as power spectral density (PSD). PSD is the complex Fourier transform multiplied by its complex conjugate and divided by the sample time. The result is a real function. The unweighted and A-weighted results are similar above about 250 Hz (remember this is a log scale plot). Only below this frequency do the two curves significantly diverge. This is reasonable since it is harder to isolate a chamber from low frequency noise than from high frequency noise. The unweighted SPL is 56.4 dB, while the A-weighted SPL is 25.4 dB.

Human perception of sound levels is a function not only of frequency, but also of level. The A-weight filter addresses the frequency dependence, but not the level dependence. An additional measure of sound called loudness was developed to account for both the frequency and the amplitude dependence. Loudness is fundamentally different from SPL in that it is a subjective measure; the mathematical descriptions of loudness are based on the subjective responses of groups of listeners to test sounds. The initial description of the loudness was proposed by Fletcher and Munson in 1933 [32]. They also defined the unit of loudness, the Phon.

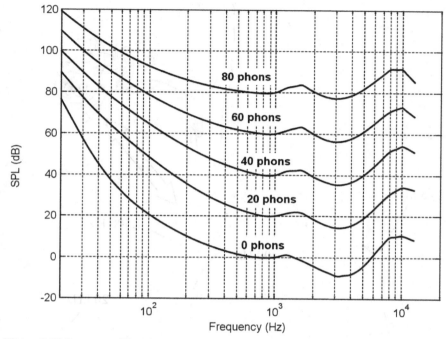

Figure 2.19 Contours of Equal Loudness Defined in ISO226

Descriptions of loudness have been revised several times since then, and the current definition is specified in ISO226 [33]. Figure 2.19 shows contours of equal loudness.

One other interesting feature of human hearing is how we perceive frequency intervals. The part of the ear that determines the frequency of a sound is the cochlea. This is a tube coiled into a spiral that serves to convert mechanical vibrations into nerve signals that can be interpreted by the brain. A tapered membrane called the basilar membrane in the cochlea vibrates in response to sound transmitted through the eardrum and the tiny bones of the middle ear. Since the basilar membrane is tapered, different frequencies excite different portions of it (Figure 2.20). Research has showed that distance to the point of maximum displacement is proportional to the logarithm of the input frequency. That suggests people are hard wired to perceive frequency intervals on a log scale. Of course, this is exactly how frequency intervals are described on a musical scale.

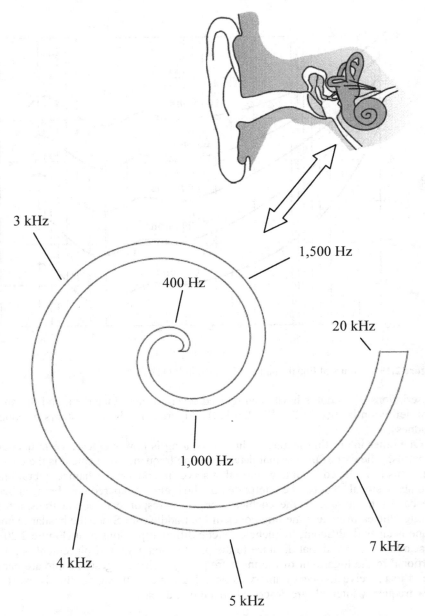

Figure 2.20 Basilar Membrane Sensitivity to Frequency (ear diagram, Wikipedia Commons, image is in the public domain)

Finally, it's worth noting that the logarithmic frequency scale is sometimes divided into discrete sections of either an octave or 1/3 octave in width. By international standard, octave bands are described in terms of the center frequencies of the bands with one band centered on 1000 Hz [34]. 1/3 octave bands are identical in concept, with the only difference being the interval spanned by each band. In musical terms, an octave band spans 12 half steps and a 1/3 octave band spans 4 half steps. Table 2.6 shows the center frequencies for octave bands and 1/3 octave bands that span the human hearing range.

Table 2.6 Octave Band and 1/3 Octave Band Center Frequencies

Octave Band Center Frequency (Hz)	1/3 Octave Band Center Frequency (Hz)
	20
	25
31.5	31.5
	40
	50
63	63
	80
	100
125	125
	160
	200
250	250
	315
	400
500	500
	630
	800
1000	1000
	1250
	1600
2000	2000
	2500
	3150
4000	4000
	5000
	6300
8000	8000
	10000
	12500
16000	16000
	20000

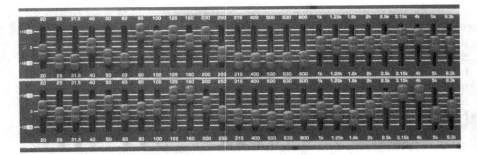

Figure 2.21 1/3 Octave Band Graphic Equalizer

A familiar place to look for octave bands and 1/3 octave bands is on graphic equalizers. Figure 2.21 shows a 31 band, 2 channel graphic equalizer. The potentiometers are clearly labeled with 1/3 octave band center frequencies.

2.6 Graphical Representations of Sound

Previous sections have described sound levels and how humans perceive sound levels. The next level of sophistication is sound quality – quantifying the subtle psychoacoustic process by which humans determine whether they like sounds. It's quite possible for two sounds to have identical levels and frequency content and yet sound completely different. Figure 2.22 shows 1/3 octave band plots of two different sounds. The two sounds have identical levels in each band (in fact, one was filtered to match the other). One plot was calculated from a segment of orchestral music and the other was calculated from a recording of a hydraulic pump.

In rough terms, psychoacoustics is the field concerned with explaining why one of the sounds represented in Figure 2.22 is more pleasing that the other. There are many objective psychoacoustic metrics in the literature, but there is no one measure of 'goodness' that works in all situations. In musical acoustics, the word timbre is often used to describe differences in sounds. Unfortunately, there is no universal definition of timbre. A 1973 ANSI Standard [35] defines timbre succinctly: 'Timbre is that attribute of auditory sensation in terms of which a listener can judge two sounds similarly presented and having the same loudness and pitch as being dissimilar'. This is essentially a subjective description in contrast to the objective definitions of loudness and pitch.

There is a further problem in that there is often not even a subjective definition of what a 'good' sound is. Reducing the problem to its most basic elements, a sound is good if it is what the listener expects. A good sounding race car engine obviously makes a very different noise than a good sounding acoustic guitar, yet both sounds can elicit an emotional response in a listener. Think of a nice small block V-8, breathing through dual quad carburetors and exhausting through headers. Then think of Willie Nelson playing his legendary classical guitar, Trigger. While both sounds might represent transcendent achievements in their own area, they have nothing else in common.

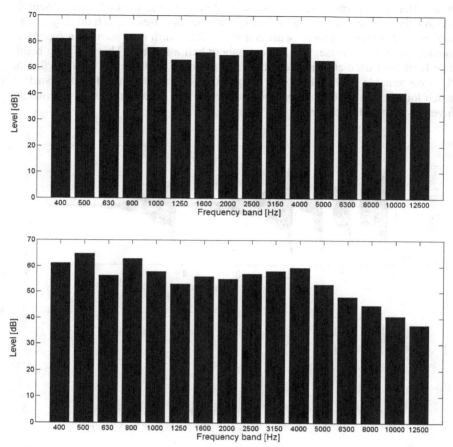

Figure 2.22 1/3 Octave Band Levels of a Hydraulic Pump (top) and A Mozart Symphony (bottom) (courtesy, Head Acoustics, www.headacoustics.com)

While there is no universally applicable psychoacoustic metric, presenting sounds in time-frequency domain is a very powerful means of analyzing and classifying them. Time domain (amplitude vs. time) shows when some acoustic event occurs. Frequency domain (amplitude vs. frequency) shows the frequency components that are present in the sound. Time-frequency domain shows not only what frequencies are present in the sound, but also when those frequencies are present [36].

Figure 2.23 shows a short, five-note guitar passage in time, frequency and time-frequency plots. The time domain plot at the top clearly shows the five notes along with their starting times, duration and decay. However, frequency information is not discernable. The second plot shows which frequencies are present in the recording, but not when they occur. The time-frequency plot (spectrogram) at the bottom shows not only what frequencies are present, but when each is present.

In addition, the spectrogram shows other interesting features that aren't discernible in either of the other two plots. The first note increases in pitch quickly before settling at a stable frequency. This is a slide. The fourth note starts at a fixed frequency and then increases in pitch over about half a second. This is a sting bend. The final note oscillates in pitch for about a second and a half. This is vibrato. Also discernable is that the lower frequencies of the 4th and 5th notes sustain much longer than do the higher frequencies and that decay time is approximately inversely related to frequency.

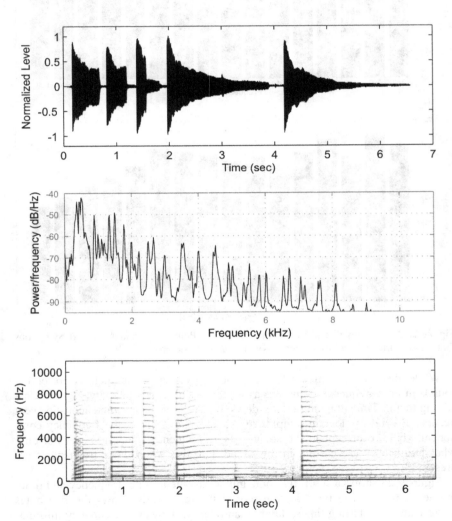

Figure 2.23 Time, Frequency and Time-Frequency Plots of a Five Note Phrase

There are many different methods of transforming time domain signals into time-frequency domain. However, it's enough here to limit ourselves to the spectrogram. The spectrogram is the most widely used time-frequency method, mostly because it is a simple extension of the fast Fourier transform (FFT) and is computationally efficient. Improved methods are available [37, 38], though at the price of increased computational time.

Figure 2.22 showed that the 1/3 octave band levels of a piece of classical music and recording of a hydraulic pump were nearly identical. Figure 2.24 shows spectrograms of the same two recordings. Now, it's clear which one is which.

Sounds from stringed instruments almost never consist of a single frequency component. Rather, they contain a fundamental frequency and harmonics – components whose frequencies are integer multiples of the fundamental frequency. These harmonics are clearly visible in the spectrogram of the five note phrase shown at the bottom of Figure 2.23. The phrase consists of five single notes, but each note clearly has many harmonics of the fundamental frequency.

Harmonics can be described using two equivalent ideas, one physical and one mathematical. The natural frequencies (in Hertz) of an ideal string are integer multiples of the fundamental frequency.

$$f_n = \frac{n\pi}{L}\sqrt{\frac{T}{\rho}} \tag{2.11}$$

Where L is length, T is tension and ρ is mass per unit length. For reasons that will be described in more detail later, plucking a string excites many modes and the resulting sound is made of the fundamental and higher natural frequencies. These higher modes are clearly visible in both frequency and time-frequency plots.

A more mathematical explanation is based on the Fourier series approximation that forms the basis of the Fourier transform. The fundamental assumption is that any periodic signal (like the sound made by a plucked string) can be approximated by a summation of sinusoidal functions.

$$f(t) = \frac{A_0}{2} + A_1 \cos \omega t + B_1 \sin \omega t + A_2 \cos 2\omega t + B_2 \sin 2\omega t + \cdots$$

$$or \tag{2.12}$$

$$f(t) = \frac{C_0}{2} + C_1 e^{i\omega t} + C_2 e^{2i\omega t} + \cdots$$

Where ω is the fundamental frequency of the periodic signal. The two expressions in Equation 2.12 are equivalent since A_n and B_n are real while C_n is complex. These coefficients define how much of each term must be added to the summation in order to reconstitute the original signal. Plotting the coefficients as a function of frequency gives the frequency domain form of the original signal.

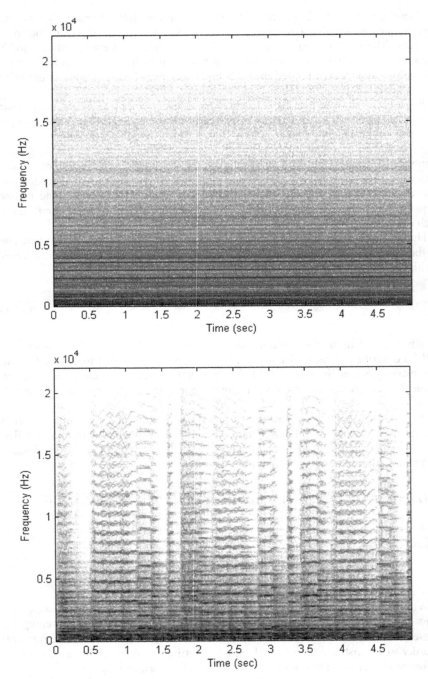

Figure 2.24 Spectrograms of Pump Noise and a Symphony

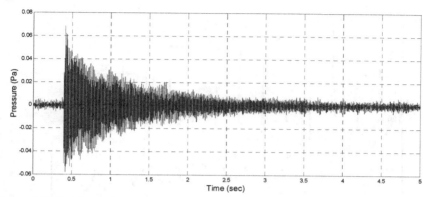

Figure 2.25 Envelope of a Plucked Note

If the original signal is sinusoidal, then there should only be one non-zero coefficient. After all, it should only take one sine wave to approximate a sine wave. However, if the original signal is not sinusoidal, additional terms are required. The frequencies of these additional terms are integer multiples of the fundamental and are, thus, harmonics.

Figure 2.25 shows the sound data for an open high E string plucked using a thin plastic pick. Since there is undesirable noise at very low frequencies and these frequencies are below the first resonant frequency of the string, an A-weight filter was applied to the data.

While the amplitude and the decay are easily determined, it's essentially impossible to see which frequencies are present in the signal. By transforming the data into frequency domain, one can clearly see that the signal is composed of many discrete frequencies and that they appear to be evenly spaced along the frequency axis as shown in Figure 2.26.

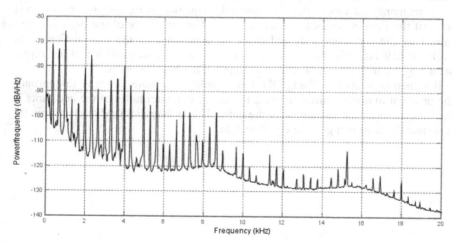

Figure 2.26 Frequency Domain Representation of Plucked Note

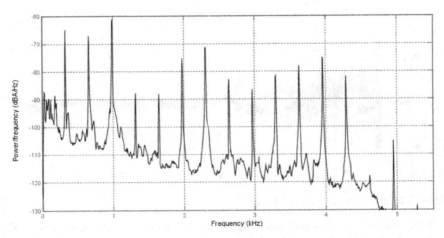

Figure 2.27 Frequency Domain Representation of Plucked Note, 0–5 kHz

It is interesting to note that Figure 2.26 shows measurable content all through the human hearing range, even though the fundamental frequency is only 329.4 Hz. Also, the amplitude of the individual peaks doesn't follow any obvious pattern. These amplitudes show the relative contribution of each frequency component, so they can be used as a tool for quantifying tonal quality. A note with lots of treble would have a larger proportion of the energy at higher frequencies. If that same note also had less content at the lower frequencies, it might be characterized by a musician as 'thin'.

By zooming in on the range from 0–5000 Hz, one can identify the fundamental frequency and verify that successive peaks are indeed integer multiples of the fundamental (Figure 2.27). They are, thus, harmonics as defined earlier.

In mathematical terms, the fundamental frequency is represented by the $n=1$ term of the Fourier series in Equation 2.12. The amplitude of the peak is determined by the constants – A_1 and B_1 or C_1 depending on the formulation. The peaks at 659 Hz, 988 Hz, 1318 Hz and so on are the harmonics and are represented by the $n=2$, $n=3$, $n=4$ and so on terms of the Fourier series.

If someone wanted to change the tonal quality of a note, the way to approach the problem in mathematical terms is to change the amplitudes of the constants in the Fourier series representation of the note. This is essentially what electronic filters (like the ones in a graphic equalizer) do.

3 Structure of the Guitar

All guitars, no matter the type, have components in common and are subject to the same types of loads. The strings must be brought to the correct tension for the instrument to be in tune, so the neck and body must resist the resulting compressive load. In addition, there are dynamic loads when the instrument is played. Finally, there are forces generated by temperature and humidity changes. While these might not be the most obvious sources of loading, they are a very common source of structural failure. It makes sense, then, to explore the structure of guitars and how they are designed to be strong enough to withstand both playing and environmental loads while still being light enough to radiate sound.

3.1 Basic Components

Most guitars have a neck and body as distinct structural components. These two components must resist the tensile forces in the strings as shown in Figure 3.1. Like all practical structures, guitars are a compromise between mutually opposing requirements. Start first with the neck. It must be strong enough to withstand the string loads while still being light enough that the instrument will balance in a comfortable way. The neck also has to have a profile that makes it comfortable to play; it can not be so wide or thick that it is difficult to grasp, nor so narrow or shallow that it is difficult to play.

Compressive Force

Figure 3.1 String Forces on Acoustic Guitar (Modified Image from Wikipedia Commons, image is in the public domain)

The neck can be considered as a cantilevered beam with an external force and an external moment as shown in Figure 3.2

R.M. French, *Engineering the Guitar*,
DOI: 10.1007/978-0-387-74369-1_3, © Springer Science+Business Media, LLC 2009

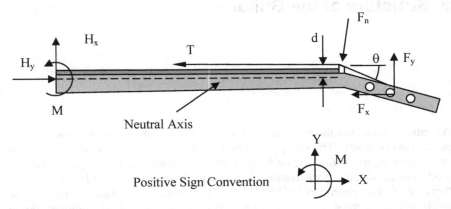

Figure 3.2 Forces on a Guitar Neck

For the neck to be in static equilibrium, the forces and moments must add to zero. Assuming the portion of the string between the nut and saddle is essentially
parallel to the fretboard, the free body diagram in Figure 3.2 can be simplified to
yield the one in Figure 3.3.

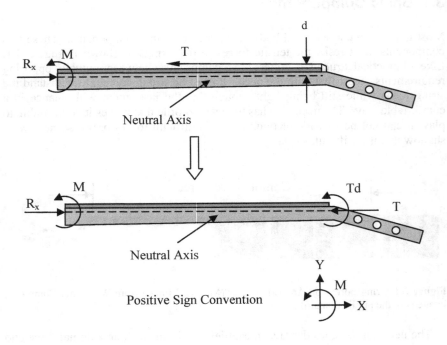

Figure 3.3 Simplified Free Body Diagram

Thus, the neck can be assumed to be subject to a compressive load equal to the sum of the string tensions, T, and to a moment equal to the sum of the string tensions multiplied by the distance from the strings to the neutral axis of the neck ($m=Td$).

The cross section of the neck cannot be accurately approximated by some simple shape like a triangle or rectangle, so it is necessary at this point to develop expressions for the centroid and the area moment of inertia of the neck.

The cross sectional shape (often called the neck profile) can be a very personal choice for players and there is certainly no one 'correct shape'. One way to account for the variation in acceptable neck profiles is to define a generic profile in terms of variable parameters. One possibility is to use one equation that approximates a wide, flat neck profile and another that approximates more of a V-shaped profile. The two functions can then be summed in various proportions to make a wide range of neck profiles. In addition, a scale factor can be added so that the resulting neck profile can be adjusted to accommodate the neck taper.

The first function is

$$y_1 = -a\left[1 - x^2\right]^{1/4} \tag{3.1}$$

Where the neck width is assumed to be 2 in. and a is the contribution of this function to the depth. If $a = 0.75$, the resulting function is shown in Figure 3.4.

The second function is

$$y_2 = b\left(x^2 - 1\right) \tag{3.2}$$

Where the neck is again assumed to be 2 in. wide and the contribution of this function to the depth is b. If $b = 0.75$, the resulting function is shown in Figure 3.5.

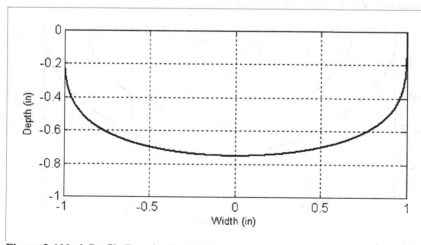

Figure 3.4 Neck Profile Function 1

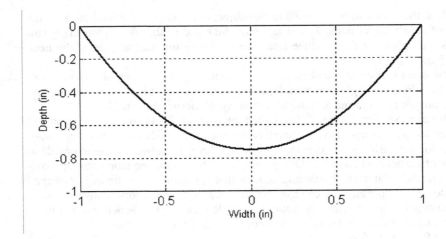

Figure 3.5 Neck Profile Function 2

The complete expression including scale factor is

$$ y = \frac{w}{2}\left[-a\left[1 - \left(\frac{2x}{w}\right)^2 \right]^{1/4} + b\left[\left(\frac{2x}{w}\right)^2 - 1 \right] \right] \qquad (3.3) $$

Where the additional parameter, w, is the desired width of the neck. The unscaled profile is 2 in. wide. If the two functions are to be combined in equal proportions and the neck is to be 0.75 in. thick, then a=0.375 and b=0.375. The resulting profile is shown in Figure 3.6. The original 2 in. wide profile is shown along with one scaled to a width of 1.5 in.

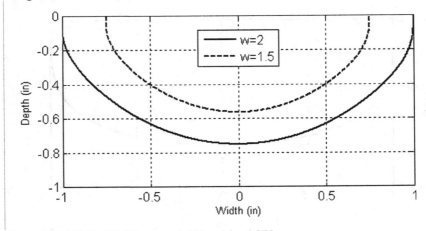

Figure 3.6 Neck Profile Where $a = 0.375$ and $b = 0.375$

The expressions that describe the deformed shape of a beam are relatively simple and can be applied to calculate the effect of string forces on the neck. It is important to note that guitar necks are not generally intended to be straight. Rather,
they generally have a slight curvature that increases the distance between the
strings and the neck. To put guitar design on a firmer analytical footing, it is useful to present a method for calculating the neck deflection.

In order to calculate the deformation of the neck as a result of string tension, it
is necessary to know the vertical location of the centroid and the area moment of
inertia of the neck cross section. The general expression for the centroid location
is

$$\bar{y} = \frac{\int y \, dA}{\int dA} \tag{3.4}$$

The denominator is just the cross sectional area and the numerator is the first
moment of area. This compact expression must be expanded in order to be evaluated

$$\bar{y} = \frac{\int_0^{w/2} y^2 \, dx}{2 \int_0^{w/2} y(x) \, dx} \tag{3.5}$$

This expression is too complicated to be evaluated analytically, though it is not
difficult when using a computational tool like Mathcad [39].

The deformation of a beam undergoing small deflections is described by a simple expression relating moment and curvature [40].

$$M(x) = E\,I(x)\frac{d^2 y}{dx^2} \tag{3.6}$$

Where M(x) is the moment along the beam, E is the elastic modulus (a material
property), I(x) is the area moment of inertia (second moment of area) along the
beam and d^2y/dx^2 is the curvature of the beam.

The second moment of area is an extension of the first moment and is calculated about the centroid

$$I = 2 \int_0^{w/2} \int_0^{y(x)} (y - \bar{y})^2 \, dy \, dx \tag{3.7}$$

This expression is evaluated numerically for all but the simplest cross-sectional
shapes. The area moment of inertia for the example neck shape is 0.0445 in^4.

Equations (3.5) and (3.7) implicitly assume that the neck is made of only one
material. This is not generally the case since most instruments use different materials for the neck and the fretboard. Additionally, most guitars (with the exception
of classical guitars) also have truss rods made of metal or graphite. In this case,

the concepts of the centroid and the area moment of inertia must be modified to account for changes in the elastic moduli of the different materials.

An easy way to account for the change in elastic modulus is to change the effective cross sectional areas of the different materials. Consider the case of the neck cross section shown in Figure 3.6. Assume the neck is made from mahogany and has a width of 2.0 in (50.8 mm). Additionally, assume there is a graphite truss rod 1/8 in (3.2 mm) wide and 3/8 in (9.5 mm) high. Finally, assume a rosewood fretboard has been added (increasing the depth of the neck). The fretboard is assumed to be ¼ in (6.4 mm) thick, but is slotted so that the effective thickness is 0.15 in (3.8 mm). If a set of calipers was used to measure the depth of the neck, the result would be 1.00 in (25.4 mm). However, the portion of the neck that can actually carry a load is 0.90 in (22.9 mm) and the following calculations will reflect this. The resulting cross section is shown in Figure 3.7.

The elastic modulus of mahogany is 1.4×10^6 psi (9.7 GPa), of rosewood is 1.9×10^6 psi (13 GPa) and of graphite is 20×10^6 psi (138 GPa). The respective effective areas can be modified so that subsequent calculations can proceed as if the elastic modulus of all the materials was the same. For this example, the modulus values will be normalized to match mahogany. Effective areas are determined by multiplying the width of the element by the ratio of elastic moduli since the area moment of inertia in this case is proportional to width. There are four areas to be calculated, three for the different elements and one for the rectangle of mahogany that is removed to so the truss rod can be installed. The material removed will be considered to have a negative area. Properties of the different components are given in Table 3.1. Note that the cross-sectional area of the fretboard in column two is 0.30 in² to reflect the fact that the unslotted depth is 0.15 in.

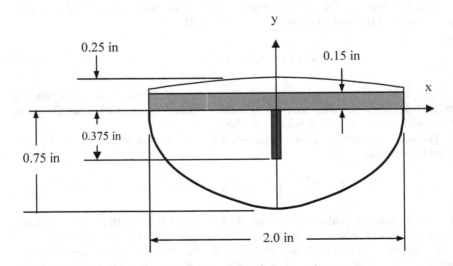

Figure 3.7 Cross-Section of Neck with Fret Board and Truss Rod

Table 3.1 Intermediate Results for Centroid Calculation

Element	Actual Area (in²)	Elastic modulus (Mpsi)	Effective Area (in²)	Vertical Centroid Location (in)
Neck	1.156	1.4	1.156	−0.312
Truss Rod Slot	−0.0469	1.4	−0.0469	−0.188
Truss Rod	0.0469	20	0.938	−0.188
Fretboard	0.30	1.9	0.57	0.075

The expression for the vertical centroid location of a shape made of multiple elements is

$$\bar{y} = \frac{\sum\limits_{i=1}^{N} y_i A_i}{\sum\limits_{i=1}^{N} A_i} \tag{3.8}$$

Where y_i is the vertical location of the centroid for element i and A_i is the effective area of element i. The centroid location for the neck cross section with the fretboard and truss rod is −0.185 in.

The expression for the area moment of inertia for a part made of multiple elements is

$$I = \sum_{i=1}^{N} I_i + A_i d_i^2 \tag{3.9}$$

Where N is the number of elements and d_i is the distance from the centroid of the complete part to the centroid of element i. Since this distance is to be squared, the sign doesn't matter and the distances are expressed here as positive numbers. A_i is the effective area and I_i is the effective area moment of inertia of element i. Table 3.2 presents these values.

Table 3.2 Intermediate Results for Area Moment of Inertia

Element	I_i (in⁴)	A_i (in²)	d_i (in)
Neck	0.0445	1.156	0.127
Truss Rod Slot	-5.49×10^{-4}	−0.0469	0.125
Truss Rod	7.85×10^{-3}	0.938	0.125
Fretboard	7.63×10^{-4}	0.57	0.260

The effective area moment of inertia for the neck, corrected for the added stiffness of the truss rod and the fretboard is 0.124 in⁴.

Since the neck is tapered, the area moment of inertia tapers along the neck. While not strictly necessary, it is helpful to develop a simple expression for *I(x)* before calculating the deformation. Assume the width at the nut is 2.00 in and the width at the 12th fret is 2.30 in. Thus, the expression for the width of the neck from the nut to the 12th fret is

$$w(x) = 2 + x/40 \tag{3.10}$$

A simple continuous expression for the area moment of inertia as a function of position along the neck can be found by fitting a polynomial to data from a few points along the neck as shown in Figure 3.8. An equivalent expression can certainly be derived directly, but this approach is a convenient shortcut that avoids lengthy symbolic manipulations.

To find the deflection along the neck due to the moment exerted by the strings, Equation (3.6) is rewritten with the curvature on the left side and known parameters on the right.

$$\frac{d^2 y}{dx^2} = \frac{M}{E\,I(x)} \tag{3.11}$$

The simplest case is the one in which the beam has a constant cross section. Then, all the terms on the right side of Equation (3.9) are constant and the deformed shape of the beam is a second order polynomial.

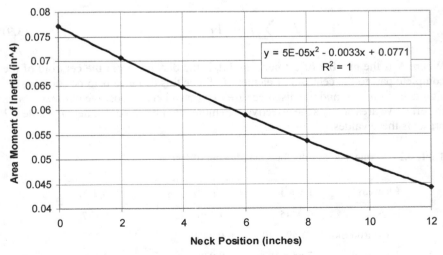

Figure 3.8 Curve Fit of Area Moment of Inertia along Neck

$$y(x) = \frac{M}{EI}x^2 + c_1 x + c_2 \tag{3.12}$$

Where c_1 and c_2 are integration constants that can be calculated by applying the boundary conditions. The neck is modeled as a cantilevered beam, so the displacement at the root is zero, $y(0) = 0$. Also, the slope at the root is zero, $y'(0) = 0$. For this particular case, both constants of integration are zero.

$$y(x) = \frac{M}{EI}x^2 \tag{3.13}$$

Assume the guitar is strung with typical nylon strings that have a cumulative tension of 85 lb. Assume also that the neck is made of mahogany with an elastic modulus of 1.690×10^6 psi (11.65 GPa). The moment is constant along the neck and the distance from the centroid to the top surface of the neck (fretboard) is 0.311 in (7.90 mm). If the strings are assumed to be 0.150 in above the fretboard at the nut, the distance from the centroid to the point of application of the tension, d, is 0.461 in and the resulting moment is 39.185 in-lb.

$$y(x) = \frac{M}{E} \iint \frac{1}{a_2 x^2 + a_1 x + a_0} \, dx \, dx \tag{3.14}$$

Where a_2, a_1 and a_0 are the coefficients of the curve fit in Figure 3.8. Evaluating Equation (3.14) and assuming zero slope and curvature at $x = 0$ (cantilevered boundary condition) gives the deflection shown in Figure 3.9. Note that solution for the tapered neck is compared to that for the mathematically much simpler case of a constant cross section. While practical guitars are not actually made with un-tapered necks, this simplifying approximation gives a surprisingly accurate result.

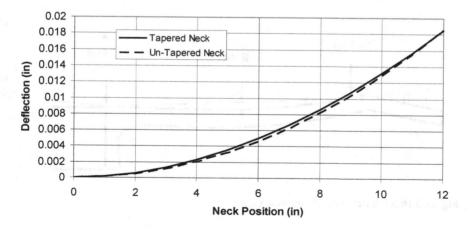

Figure 3.9 Deflection along Neck of Classical Guitar Due to String Tension

Calculating neck deflection is of little practical use without a clear idea of what an acceptable design might be. The first thing to note is that it is not generally desirable for the neck to be perfectly flat. Rather, a slight upward curvature (known as relief) is intentionally built into most guitars. There is no one correct value for relief since it can depend on the height of strings above the soundboard, the style of music being played and the type of guitar.

A simple way to measure neck relief is to place a capo (a clamp that stops the strings wherever it is placed on the neck) on the first fret and to fret a selected string where the neck joins the body. For steel string acoustic guitars, this is usually the 14th fret and for classical guitars, this is usually the 12th fret. Since the string is straight between the two frets, relief is apparent as space between the strings and the frets between these two points (as shown Figure 3.10). Measured this way, neck relief in the range of 0.010 – 0.030 in (0.25 – 0.75 mm) is typical.

Wood is sensitive to changes in both temperature and humidity. Furthermore, wood can creep under load – permanently deform even though the stresses are lower than the yield stress. Thus, the neck relief of a guitar can change over time and may need to be adjusted periodically.

With the exception of classical guitars, most guitars have a truss rod to stiffen the neck, allow some adjustment of the neck deformation or both. The different designs can be roughly grouped into three categories: Fixed (non-adjustable), Tension and Double Acting. Fixed truss rods are simply reinforcing elements in the neck that increase the bending stiffness. Current practice is to use unidirectional carbon fiber, though steel rods are still sometimes used. Figure 3.11 shows the placement of a fixed truss rod in the neck. Even a relatively small fixed truss rod can have a major effect on the stiffness of the neck because of the differences in elastic modulus of the materials. For example, the elastic modulus of wood is on the order of 1.7×10^6 psi while the elastic modulus of steel is 29×10^6 psi. The elastic modulus of a unidirectional carbon fiber beam depends on the proportions of fiber and matrix, however, a typical value is 20×10^6 psi.

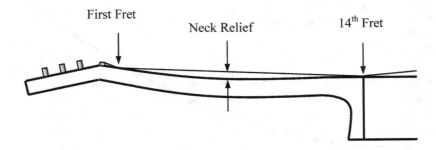

Figure 3.10 Neck Relief on an Acoustic Guitar

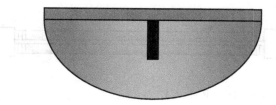

Figure 3.11 Placement of Fixed Truss Rod

A tension rod generates an internal force in the neck to counter the deformation caused by string tension. Early tension rods were straight, but had to be placed below the elastic axis of the neck. Tightening a nut on one end places the rod in tension, thus, generating both a compressive force and a moment as shown in Figure 3.12. Straight rods were first used in acoustic guitars with fixed necks (glued on and not easily removed), so adjustments were made at the head end. That end was generally accessed though a slot in the top of the headstock, near the nut.

Generally, tension rods are installed with a slight dip in the middle, increasing the resulting vertical force and making them more effective. These curved rods are routinely installed in both electric and acoustic guitars. The end at which the rod is adjusted depends on how it is installed. Guitars with the adjustment at the head end often have a small, removable cover over the end of the truss rod. Whether straight or curved, though, a tension rod can only induce convex curvature along the fretboard.

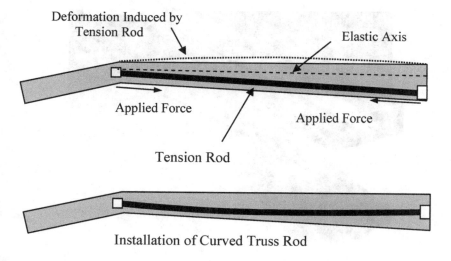

Figure 3.12 Placement of a Tension Rod in the Neck

Figure 3.13 Double Acting Truss Rod

The most versatile type of truss rod is double acting rod. While designs differ, a representative one consists of two circular rods with threaded ends and mounted into end blocks as shown in Figure 3.13. Turning the adjustment nut at the left forces the bottom of the blocks apart or together (depending on the direction of rotation) and, thus, causes the rod to arch either up or down. Some double acting truss rods are wrapped with fiberglass tape to keep the two rods from separating in the center. Figure 3.14 shows typical installations of a double acting truss rod. Note, this design does not require a tape wrap since the end blocks apply a moment to the two rods, causing them both to bend.

The soundboard is clearly a very important structural element in any acoustic guitar. Unfortunately, it is very difficult to analyze. There is a plate equation that describes the static deformations of an isotropic plate with known boundary conditions in a compact form. However, it is a fourth order partial differential equation; it has been solved analytically only for simple shapes like rectangles and circles and for simple boundary conditions.

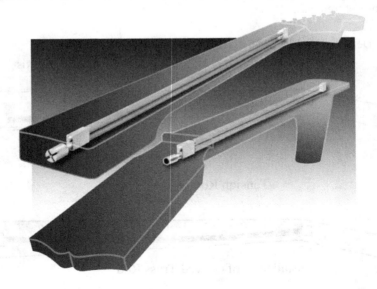

Figure 3.14 Installations of a Double Acting Truss Rod (image courtesy Stewart-MacDonald, www.stewmac.com)

A typical soundboard includes structural features that cannot be accounted for using the plate equation; wood is orthotropic rather than isotropic, the shape of the soundboard is too complex to be described by a simple algebraic function, and there are numerous braces and reinforcing plates. Thus, a mathematical representation of a guitar top simple enough to be solved analytically cannot capture the details of the actual structure.

Analyses of guitar tops appearing in the literature often use one of two approaches. The first is to make a simple analytical model and attempt to extend the results to an actual guitar top by analogy. The second is to make a discretized model of the structure, usually a finite element model [41]. A good finite element model can easily have thousands of degrees of freedom, particularly if it includes the air volume in and around the instrument. Mathematical modeling for soundboards will be discussed in more detail in Chapter 5.

The body of an acoustic guitar is constrained by the same types of opposing requirements as the neck. The combined string forces are generally 60 lb – 120 lb (267 N – 534 N) and the top must be able to withstand this force while retaining enough flexibility to vibrate in response to the strings. Because of this conflict between strength and light weight, the static string tension is large enough to noticeably deform the soundboard of an acoustic guitar.

For this discussion, a model made using thousands of degrees of freedom is not necessary for a basic understanding of how string forces deform the top plate. Indeed, a useful qualitative description can result from treating the plate as a pinned beam as shown in Figure 3.15.

The in-plane force does not cause out of plane deformation, so the beam equation can be solved using the moment as the only applied load. The resulting load-shear-moment diagram is shown in Figure 3.16.

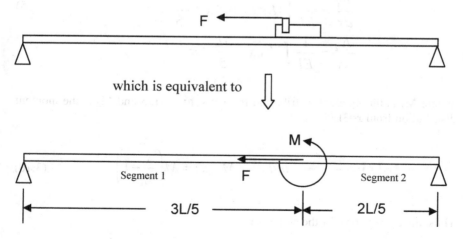

Figure 3.15 Pinned Beam Analogy to Top Deformation

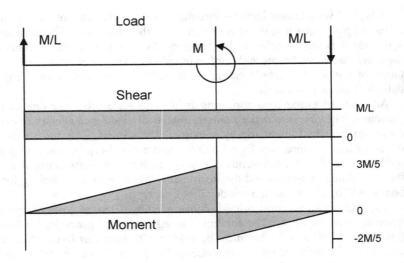

Figure 3.16 Load-Shear-Moment Diagram for Pinned Beam Analogy

Since there is a discontinuity in the moment distribution along the beam, the integral takes place in two parts.

$$\frac{d^2y}{dx^2} = \frac{M}{EI}$$

$$\frac{dy_1}{dx} = \frac{1}{EI}\int M_1 dx \quad 0 \le x \le \frac{3L}{5}$$

$$\frac{dy_2}{dx} = \frac{1}{EI}\int M_2 dx \quad \frac{3L}{5} \le x \le L$$

(3.15)

Where M_1 is the moment distribution from $x=0$ to $x=3L/5$ and M_2 is the moment distribution from $x=3L/5$ to $x=L$.

$$M_1(x) = \frac{Mx}{L} \quad and \quad M_2(x) = M\left(\frac{x}{L} - 1\right)$$

(3.16)

Thus, the expressions for the slope are

$$\frac{dy_1}{dx} = \frac{1}{EI} \int \frac{Mx}{L} dx = \frac{M}{2EIL} x^2 + c_1 \quad 0 \le x \le \frac{3L}{5}$$

$$\frac{dy_2}{dx} = \frac{1}{EI} \int M\left(\frac{x}{L} - 1\right) dx = \frac{M}{EI}\left(\frac{x^2}{2L} - x\right) + c_2 \quad \frac{3L}{5} \le x \le L \tag{3.17}$$

And the expression for the displacement is

$$y_1(x) = \int \frac{M}{2EIL} x^2 + c_1 \, dx = \frac{M}{6EIL} x^3 + c_1 x + c_3 \quad 0 \le x \le \frac{3L}{5}$$

$$y_2(x) = \int \frac{M}{EI}\left(\frac{x^2}{2L} - x\right) + c_2 \, dx = \frac{M}{EI}\left(\frac{x^3}{6L} - \frac{x^2}{2}\right) + c_2 x + c_4 \quad \frac{3L}{5} \le x \le L \tag{3.18}$$

Finally, the four integration constants are identified by applying four boundary conditions

$$y_1(0) = 0$$
$$y_2(L) = 0$$
$$y_1(3L/5) = y_2(3L/5) \tag{3.19}$$
$$y_1'(3L/5) = y_2'(3L/5)$$

After applying these boundary conditions, the expressions for the displacement along the beam are

$$y(x) = \frac{M}{EI}\left(\frac{x^3}{6L} - \frac{x^2}{2}\right) - \frac{13ML}{150EI} \qquad 0 \le x \le \frac{3L}{5}$$

$$y(x) = \frac{M}{EI}\left(\frac{x^3}{6L} - \frac{x^2}{2}\right) + \frac{77MLx}{150EI} - \frac{9ML^2}{50EI} \qquad \frac{3L}{5} \le x \le L \tag{3.20}$$

The resulting displacement is shown in Figure 3.17. The solid line shows the deformed shape assuming the stiffness is constant along the beam and the dotted line shows the effect of doubling the stiffness on the left side of the applied moment (the left side of the bridge). It is typical for acoustic guitars to have heavy bracing near the sound hole and around the neck interface. Thus the stiffness of the top is significantly higher in the lower bout than the upper. When this change is made, the area to the right of the moment has positive deflection – the same behavior as generally seen in acoustic instruments.

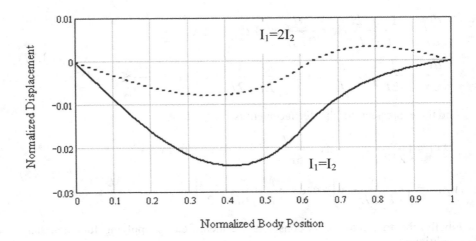

Figure 3.17 Analogous Beam Deflection

3.1.1 Classical Guitar

The classical guitar is the first of the modern types of guitar to be developed, having achieved roughly its current form by the late 1800's. One significant exception, of course, is the strings. The central feature of the modern classical guitar is the nylon strings while earlier ones were fitted with gut strings. Because of the material properties of nylon, the strings need to be brought to about 80 lb (356 N) tension for the instrument to be at standard tuning [42].

The instrument consists basically of the body and the neck. Figure 3.18 shows a representative classical guitar.

Guitars are regularly made from a range of different woods, but classical guitars have traditionally used spruce for the top and rosewood for the back and sides. Initially, tops were made from European species of spruce; now, Engelmann and Sitka spruce are widely used. The neck is often mahogany and the fretboard is ebony. There is a decorative rosette around the soundhole that is usually made from very small tiles of dyed wood. The bridge is usually made of ebony or rosewood. Figure 3.19 shows the rosette on a classical guitar made by Samo Šali.

The load-carrying structure of the classical guitar is generally very simple, though it is quite refined. There is no single accepted design for classical guitars as there is for violins, so some representative design has to be used. The description here is of a traditional instrument established by Torres and others in the 19th century.

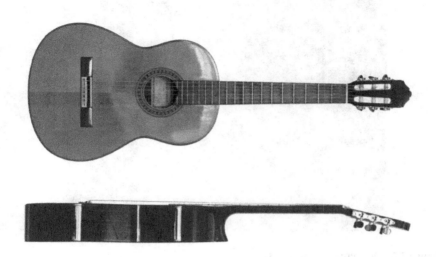

Figure 3.18 Classical Guitar (Wikipedia Commons – image is in the public domain)

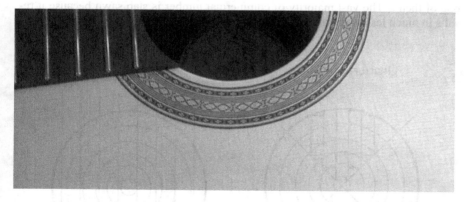

Figure 3.19 Soundhole Rosette on a Classical Guitar (image courtesy Samo Šali, http://salilab.org/samo/web6/guitars.htm)

The top is made of book-matched, quarter-sawn spruce or, less often, cedar. Quarter-sawn wood is cut from the log so that the grain is perpendicular to the face of the resulting boards. The name comes from the fact that the log is split into quarters before being sawn into planks. Figure 3.20 shows collection of short log sections that have been split prior to being sawn into planks.

Figure 3.20 Quartered Log Sections

Figure 3.21 shows how the log quarters can be sawn into planks. Note that there are several common sawing patterns for quarter sawing logs and this is just one of them. The vast majority of commercial lumber is slab sawn because it results in much less waste than does quarter sawing.

Quarter-Sawn Log   Slab-Sawn Log

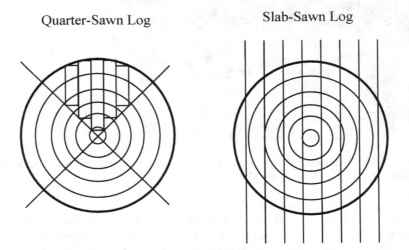

Figure 3.21 Quarter-Sawn and Slab Sawn Logs

Slab-sawn lumber is not preferred for instrument making for a number of reasons. From a structural standpoint slab-sawn boards are problematic because the grain lines tend to straighten out as the wood dries and the boards tend to cup. Instruments made from quarter-sawn lumber tend to be more dimensionally stable [43]. There are aesthetic reasons as well; the grain of quarter-sawn wood tends to be even and sometimes displays subtle, attractive effects. Finally, centuries of tradition have conditioned musicians to associate quality instruments with quarter-sawn wood. Figure 3.22 shows a quarter-sawn board. Note that the center board from a slab sawn log has vertical grain just like a quarter sawn board.

Guitar tops and backs are generally book matched. A book matched board is one that is cut in half and joined to make a plank that is twice as wide and half as thick. Figure 3.23 shows the process for making a book matched plank. Book matching results in a board whose grain is symmetric about the centerline and can be done for aesthetic or structural reasons. The guitar back in Figure 3.24 is made from book-matched rosewood with a tapered center strip of lighter wood.

Figure 3.22 A Quarter-Sawn Board

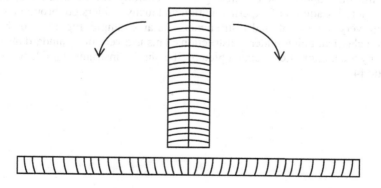

Figure 3.23 Making A Book Matched Board

Figure 3.24 Book Matched Back with Light Center Insert (image by the author, reproduced courtesy of Taylor Guitars, www.taylorguitars.com)

The body shape of classical guitars is fairly standardized. Richard Bruné [44] has laid out recommended proportions for the classical guitar as shown in Figure 3.25. Indeed, the vast majority of classical guitars have proportions similar to these. Note that these are just proportions and not absolute measurements. The length of the body is 1 and all other dimensions are to scale. If, for example, the body is to be 21 in (533 mm) long, simply multiply all the numbers by 21 in.

While the majority of classical guitars generally follow traditional designs, there is a rich tradition of experimentation and many luthiers go through a stage of making very unconventional instruments. As an example, Figure 3.26 shows an archtop classical guitar under construction. This is a very non-standard instrument made by the author after seeing a picture of a similar instrument made by Bob Benedetto [45].

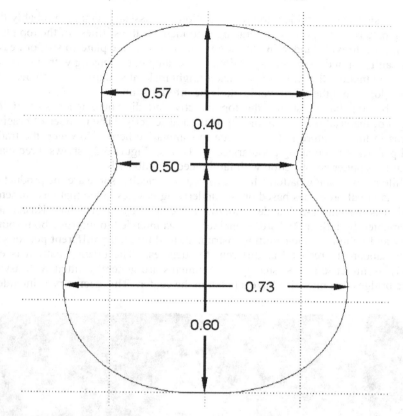

Figure 3.25 Recommended Proportions for the Classical Guitar

Figure 3.26 An Archtop Classical Guitar under Construction

The design feature subject to most variation in classical guitars is probably the bracing pattern. The purpose of bracing is to increase the stiffness of the top plate enough for it to withstand string tension without making the plate so stiff or heavy that it can't respond to the string motion. For example, dispensing with the braces and simply making the entire plate thicker might make it strong enough to bear the in-plane loads, but the result would likely be inferior sound.

The bracing has to stiffen the top directly and distribute loads so that the strength of material making up the top is used efficiently. The number of bracing patterns in the literature is far too large to summarize here. However, the traditional pattern is fan bracing popularized by Torres. Figure 3.27 shows three classical guitars under construction with fan-braced soundboards.

While many bracing patterns have evolved empirically, some are the product of more analytical processes based on the underlying physics. One that has attracted attention is the Kasha-Schneider pattern. Developed by Dr. Michael Kasha and implemented by luthier Richard Schneider, it was intended to improve both sound quality and volume of the resulting instrument [46] by tuning different portions of the soundboard to respond at different frequencies. The bracing pattern is extremely asymmetric and Kasha style instruments are generally fitted with asymmetric bridges that may also have a notch in the center. This geometry is intended

Figure 3.27 Three Partially Completed Guitars with Fan-Braced Soundboards (image courtesy John Elshaw, www.elshawguitars.com)

to match the bridge to the mechanical impedance of the soundboard. Kasha-braced guitars often feature an offset soundhole, though it is important to note that offset soundholes have been used on more conventional designs as well in an effort to generate a larger radiating surface on the guitar top. Figure 3.28 shows a Kasha style soundboard with an offset soundhole. This bracing pattern has not been widely accepted, though a number of successful luthiers use it and attest to its superior sound quality.

An extremely unconventional bracing pattern that has also resulted from analytical arguments is used by Samo Šali, an accomplished luthier and engineer working in Slovenia. He has proposed a method for determining brace geometry that requires curved braces [47] and makes instruments using this system. Figure 3.29 shows the braces on one of his instruments before the back is glued on.

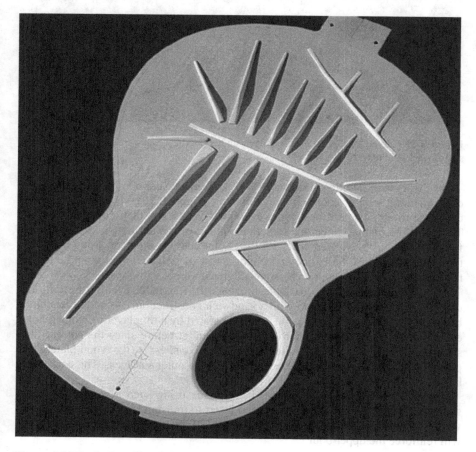

Figure 3.28 Kasha Bracing (image courtesy Guild of American Luthiers, www.luth.org)

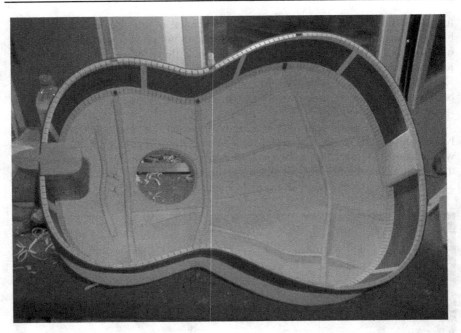

Figure 3.29 Curved Braces on a Šali Guitar (image courtesy Samo Šali, http://salilab.org/samo/web6/guitars.htm)

As this is being written (December 2007), an increasing number of luthiers are experimenting with or adopting lattice bracing (Figure 3.30). The lattice tends to distribute the bracing stiffness more evenly across the top than does fan bracing. The elements of the lattice are sometimes made of wood, though graphite reinforced wood has been used successfully. Some luthiers are even using balsa for the wood portion of the lattice and gluing strips of unidirectional graphite to the top of the braces. This approach can result in extremely light tops.

If the luthier wishes to distribute bracing stiffness more evenly, the limiting case would be to make the top with a sandwich construction. Some luthiers have been successful in using thin wood plates separated by a thin layer of Nomex honeycomb [48]. This can be a particularly efficient structure since the parts of the structure that can withstand tensile and compressive stresses (the wood plates) are placed where those stresses are the highest – at the outer surfaces. The core layer basically carries shear loads.

Figure 3.31 shows a sound board being made using such a core layer. The soundboard shown here will have a wood insert to transmit the bridge loads to the wood plates. The insert also will form the edge of the soundhole after it is cut and will reinforce the upper bout.

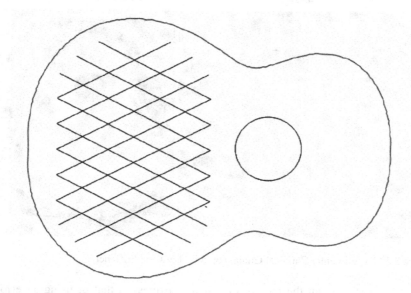

Figure 3.30 Schematic of Lattice Bracing on a Classical Guitar Top

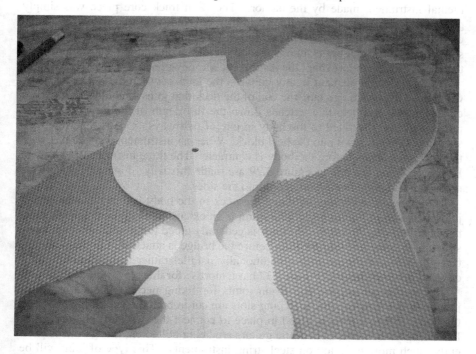

Figure 3.31 A Soundboard Being Made With a Nomex Honeycomb Center Layer (image courtesy Reynolds Guitars, www.reynoldsguitars.com)

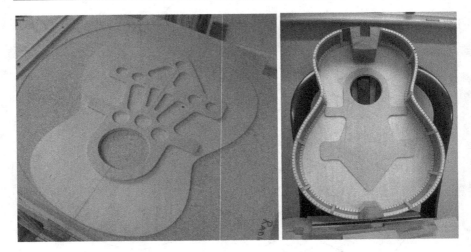

Figure 3.32 Experimental Classical Guitar Top with Plywood Stiffener

Another variation on the idea of distributed stiffness is that of using a perforated sheet of light birch plywood as the core layer. Figure 3.32 shows an experimental instrument made by the author. The ¼ in thick core piece was simply glued to the inside of a plywood top and then covered with a layer of 1/16 in aircraft grade birch plywood. This design is probably not as efficient as the one shown in Figure 3.31, but is more suited to the skills of undergraduate engineering technology students for whom the instrument was developed.

A key structural element in any guitar is the joint between the neck and body. On almost all acoustic guitars, the end of the neck that joins the body has a curved extension called a heel that extends down the full depth of the body. Classical necks are generally fixed to the body in one of two ways. One is to fit the bent sides into notches sawn into the heel block. When an instrument is made this way, the neck is attached before the body is complete. The three instruments shown in Figure 3.27 and the one in Figure 3.29 are made this way. Figure 3.33 shows a notched heel before being assembled into the sides.

The other common way of fixing the neck to the body is a mortise and tenon joint as shown in Figure 3.34. The tenon is an extension of the heel and the mortise is milled into the body. The neck is generally glued to the body after the body structure is complete, though often before the bridge is attached. On classical guitars, the sides of the tenon are traditionally parallel rather than tapered as in a dovetail joint; the body in Figure 3.32 has a mortise for this type of neck joint.

A variation on mortise and tenon joints for fixing necks to bodies is spline joints. In this type of joint, matching slots are cut in both the end of the neck and the body. Then, a plate is glued in place to connect the two parts (Figure 3.34). Increasingly, the necks of acoustic guitars are being bolted on, though this is presently much more common on steel string instruments. This type of joint will be described in detail in a subsequent section.

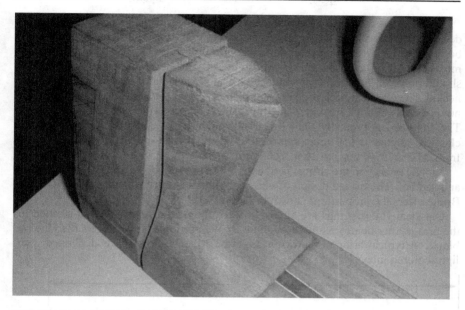

Figure 3.33 A Neck Blank with Notched Heel (image courtesy Samo Šali, http://salilab.org/samo/web6/guitars.htm)

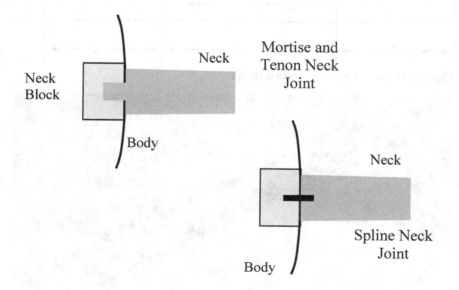

Figure 3.34 Two Methods for Connecting Guitar Necks

The neck of the classical guitar is generally made of mahogany and typically does not include a truss rod or any other type of reinforcement. The neck joins the body at the 12th fret, so the cantilevered portion of the neck is shorter than that of most steel string acoustic guitars whose necks join the body at the 14th fret. It should be noted however that some steel string instruments are made with 12 fret necks. The classical neck is also wider than is typical for a steel string guitar. The nut is usually about 2 in. wide (50.8 mm) and can be as wide as 2.125 in. (54 mm). The fretboard is typically flat or nearly so. The most common scale length for classical guitars is 650 mm (approx 25.5 in) and scale lengths commonly range from 25.2 in. (640 mm) to 26 in. (660 mm).

Kinetic energy in the strings is transferred to the soundboard through the bridge and saddle. The bridge design of classical guitars is fairly standardized. Richard Bruné offers the proportions shown in Figure 3.35.

Classical guitar strings are almost universally made from nylon and are tied to the bridge after being passed through holes in a portion of the bridge called the tie block. A typical bridge is shown in Figure 3.36. The tie block is decorated with light-colored inlays.

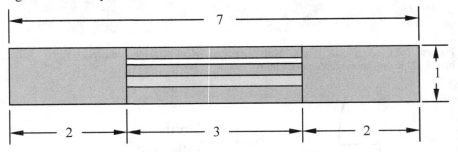

Figure 3.35 Proportions for a Classical Bridge

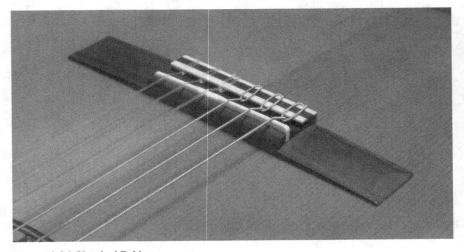

Figure 3.36 Classical Bridge

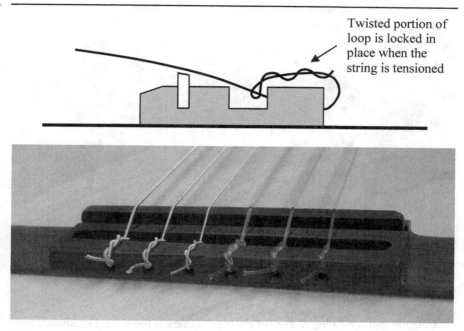

Twisted portion of loop is locked in place when the string is tensioned

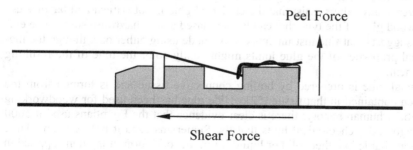

Figure 3.37 Classical Strings Tied to Bridge

Since a classical guitar can drift out of tune if the knot at the bridge slips, one of a few specific techniques is used to tie the strings to the bridge. A common method is shown in Figure 3.37.

The bridge is simply glued to the top of the soundboard in the appropriate position. The total tension in the six strings is on the order of 85 lb (378 N). If the glue area is 7 in^2 (4515 mm^2), the shear stress on the glue joint is 12.1 lb/in^2 (83.4 kPa). This is far below the yield strength in shear of even weak glue [49, 50, 51]. Unfortunately, the glue layer under the bridge is not exclusively in shear; the rear of the bridge applies a tensile load as well. Care must be taken to ensure that the peel strength of the glue is not exceeded. Forces exerted by the bridge on the glue layer are shown in Figure 3.38.

Peel Force

Shear Force

Figure 3.38 Forces Exerted by the Bridge on the Glue Layer

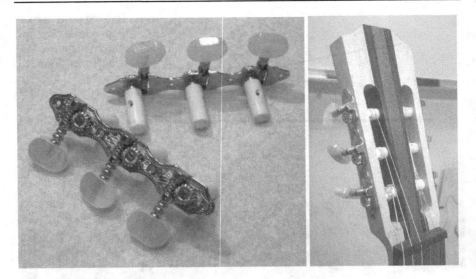

Figure 3.39 Classical Guitar Tuners

A clearly distinguishing feature of classical guitars is the slotted headstock and tuners with large diameter rollers. The tuners are almost universally mounted three to a plate with one plate for each side of the headstock. The roller spacing is typically 35 mm (1.378 in) and the roller diameter is typically 10 mm (0.394 in). Figure 3.39 shows a set of well-made classical tuners.

This chapter has focused on the components that make up the structure of the guitar, but the discussion is not complete without considering the glue that holds the structure together and the finish that protects it.

Guitars can be made, and probably have been made, using just about any substance that can join wood with acceptable strength. There are many good glues currently available, but the two most popular choices among luthiers are probably animal hide glue and Polyvinyl Acetate (PVA) wood glue. The choice of glue has been the subject of much debate and there is no sign of any emerging consensus. Some luthiers insist that fine guitars can only be made using animal hide glue. However, many others insist that the choice of glue is not critical and happily use PVA wood glue of the type that can be purchased at any hardware store. The evidence suggests that fine instruments can be made using either one, though the mechanical properties of the glue joints might subtly color the tone of the resulting instrument.

Animal glue is prepared by boiling connective tissue and is formed from the collagen contained in that tissue [52]. Hide glue has been used for woodworking throughout human history; there is clear evidence that the Egyptians used it 5000 years ago and archeological hints that the Sumerians used it before them. Hide glue is available in either solid or liquid form. In solid form, it is often supplied in granules that can be stored for extended periods before being prepared for use. The granules are soaked in water, heated and applied warm, usually around 140 °F (60 °C).

The liquid form of hide glue, while convenient to use, contains additives that keep it from gelling in the bottle [53]. In static tests, liquid hide glue has been shown to be as strong as hot hide glue when the relative humidity is less than 50%. However, the shear strength drops off rapidly at higher humidity. As a result, liquid hide glue has been universally shunned by luthiers.

Two characteristics of hide glue often referenced by its fans are that it does not creep and that it is reversible. PVA can creep under high loads, causing ridges at glued seams and possibly allowing permanent distortion of the glued structure. There are many examples of violins assembled with hide glue in the 17th century that are still being played regularly. Hide glue is also brittle and a joint can sometimes be separated with a sharp blow. In the hands of a skilled luthier, this characteristic can be a distinct advantage since instruments can be disassembled without being damaged.

Hide glue joints can be separated readily by applying heat and moisture. When re-heated, the glue returns to its liquid state, so instruments assembled with it can be disassembled as needed. This is particularly important for instruments in the violin family that are intended to be disassembled regularly for maintenance and repair. Hide glue also sticks to itself so the two surfaces to be glued can be coated with hot hide glue and allowed to cool. The joint can be formed later by bringing the two surfaces together and adding heat. It should be noted that this is a specialized technique practiced by some luthiers and is not the usual method of forming glued joints.

An argument against hide glue is that it is not convenient to use. Once mixed with water, it has a short shelf life and manufacturers generally recommend that it be used within a day or so. Since it must be applied hot, there has to be some heating vessel on the luthier's bench. Also, the working time is short, so the parts must be brought together quickly after the glue is applied.

Synthetic PVA is probably more widely used and is commercially available in several forms. It is water-based and dries as the water evaporates or is soaked into the wood. The most familiar PVA is the white glue (e.g. Elmer's) that is an indispensable part of any child's craft box. White PVA is storable, non-toxic, easy to apply and is strong enough to be used for making instruments. However, it is relatively soft when dry and weakens when exposed to heat and humidity. Yellow wood glue (e.g. Titebond) is the usual choice among luthiers choosing to use PVA. It retains the advantages of white PVA while drying to a harder and more sandable state. More recent versions are also water resistant enough to be suitable for exterior use.

The characteristics of PVA make it popular in manufacturing operations. For example, Taylor Guitars uses PVA for all their instruments. For most of the history of the company, they used a common brand of PVA wood glue. They have only recently switched to a formulation that includes an ultraviolet dye so that excess glue can be located easily under a black light.

Some other types of glue have been adopted by many luthiers. Cyanoacrylate (super glue) is often used for repairs, for applying inlays and other specialized applications. The task of bonding synthetic binding to wood body structures usually falls to a vinyl or acetone based glue (e.g. Duco Cement). Epoxy is also used for instruments, particularly those with graphite or other composite components. An epoxy joint can be disassembled only with the greatest effort, but it is not likely to creep under load as PVA glues can do. Finally, polyurethane glue is becoming more readily available and is increasingly being used to make bonds between wood components.

The final component of the classical guitar is the finish. While the finish is not part of the load-carrying structure, it is very important for several reasons. Bare wood is easily damaged. Dirt and oil from the hands of the player also tend to be rubbed into the grain of the wood and is nearly impossible to remove without damaging the instrument. The role of finish is basically to protect the wood and to make the instrument more attractive. Guitars don't commonly have finish applied to the inside surfaces, but this is occasionally done [45], presumably to keep the bare interior surfaces from absorbing humidity from the air.

The finish does affect the tone of the instrument and there is much discussion in the literature about what type of finish is preferable. Many different types of finish are used and it can be difficult to make sense of the different techniques. The most fundamental distinction is whether a chemical reaction takes place as the finish cures.

In finishes where there is no chemical reaction, a solid material is dissolved in some liquid so that the resulting mixture can be brushed or sprayed on the instrument. When the solvent evaporates, a thin, even coating of the original solid is left. The classic example is shellac.

Shellac is made by dissolving a brittle secretion of the lac bug (*Coccus lacca*) in alcohol [54]. After the liquid shellac is applied, the alcohol evaporates, leaving a thin protective layer on the instrument. The solid shellac is harvested from forests in Southeast Asia where it is found deposited on trees. In this raw form, it is called seedlac. After being refined, it is often sold as solid flakes that can be dissolved in an appropriate solvent as needed. Figure 3.40 shows shellac flakes.

Shellac has some interesting properties that have allowed it to be used in some surprising applications. When properly prepared, it is edible and has been used as a coating on pills and candy. In its solid form, it is thermoplastic and has been used for applications in which plastic would now be used. For example, the first phonograph records were pressed from shellac.

Finally, shellac is compatible with many other types of finish. An experienced violin maker once told me his rule of thumb was that any finish that will stick to wood will stick to shellac. Because of this, it is often used as a sealer or primer coat on bare wood. It is often used as a barrier to prevent stain or pigment applied to the bare wood from bleeding into the final finish.

Figure 3.40 Shellac Flakes

Shellac is generally brushed on or applied with a soft pad. Spraying shellac can be difficult since the alcohol tends to evaporate quickly from the atomized liquid. Fine classical guitars are often finished using a process called French polishing. This uses a soft cotton pad onto which a small amount of shellac is dripped. The pad is rubbed on the instrument to deposit a very thin coat of finish. The result can be a fine, thin, even finish. It is not very durable, but adds very little weight and induces little or no change in the mechanical properties of the instrument. Classical guitar makers generally agree that the thinnest, lightest possible finish results in the best sound quality. As a practical compromise, classical guitars sometimes have French polished soundboards while the rest of the instrument has some heavier, more durable finish. The alcohol solvent in shellac is flammable and can be toxic if methanol or denatured alcohol is used. Some luthiers make shellac exclusively with ethanol.

Another type of finish that cures through evaporation of a solvent is lacquer. Lacquer, in various forms, has been in existence for centuries and was used heavily in Asia. The modern form of lacquer used for musical instruments is nitrocellulose lacquer. It is made with nitrocellulose resin formed by nitrating cotton or other materials containing cellulose. The resin along with other resins and plasticizers is dissolved in a volatile solvent and the resulting mixture is applied by spraying or brushing.

Lacquer dries quickly and can be polished to a form a glossy finish. When a new coat of lacquer is applied over a previous layer, the solvent partially dissolves the previous coat. Thus, lacquer finishes are easy to repair if scratched or chipped and the repaired area is nearly impossible to distinguish from the original finish.

A significant disadvantage of lacquer is that the solvents are toxic and flammable. Fires in lacquer spraying booths are not unknown and can be quite nasty affairs when they occur. The solvent may include naptha, xylene, toluene, ketones and acetone, so exposure poses health risks. Cartridge respirators and fans with explosion-proof motors are a requirement.

A completely different class of finishes consists of compounds that undergo a chemical reaction as they cure. The simplest of these are natural oils such a tung oil and linseed oil. Linseed oil, in particular, is commonly used as a wood finish. Boiled linseed oil is generally used since the heating process polymerizes the oil, thickens it and shortens the drying time [55]. Raw linseed oil is edible and sometime used as a dietary supplement (labeled flaxseed oil since it is extracted from the dried seeds of the flax plant). Products labeled as boiled linseed oil may contain petroleum solvents and metallic drying agents and are, thus, toxic.

Linseed oil cures through oxidation, a chemical reaction. It generally soaks into the wood before completely curing, so it hardens the outer layer of wood and makes it more durable. When dry, it leaves a smooth, satin finish. A thick finish can be built up, but this can be a slow process. Tung oil is extracted from a different plant – the nut from a tung tree – but behaves like linseed oil. It differs in that it is generally thicker and the cured finish is much more water resistant. Other plant oils such as walnut oil are also used as wood finishes and may be suitable for guitars.

Traditionally, finishing agents like shellac and linseed oil have been combined with other substances to form mixtures that go by the generic name of varnish. Violin makers, in particular, draw on a long list of tree resins and other materials that can be added to either shellac or linseed oil to alter the properties of the cured finish. Shellac/alcohol based varnishes are generally called spirit varnishes. Those based on drying oils are, predictably, called oil varnishes.

Two other classes of finish are popular enough to be worth mentioning here, epoxy and finishes that rely on a catalyzing agent. Two part epoxy, if thin enough, can be a very serviceable finish. There are two part finishing epoxies on the market (e.g. ZPoxy). They are storable, easy to mix and cure to a hard finish that is impervious to water. They are too thick to be brushed or sprayed on and can be spread on with a flexible spatula.

Catalyzed finishes are probably more widely used than epoxy. Such finishes do not cure until some additional agent is applied. Catalyzed varnishes typically use a chemical agent that is added by the user before application. Such finishes tend to be durable and easy to apply in a production setting where extended drying times are problematic.

An increasingly popular finish used in production operations is a synthetic resin that cures when exposed to ultraviolet light. This is perhaps not strictly a catalyzed finish in the sense that there is no additional component added to the finish before application. However, the finish will only cure when exposed to UV light. This system is used by Taylor Guitars with very good results. The thick, uncured finish is sprayed on by an industrial robot. The sprayed instrument is then placed in a box in which ultraviolet lights are mounted. The instrument stays in the light box for less than a minute and the finish is cured when it emerges. The cured finish can then be buffed to a durable glossy finish.

3.1.2 Steel String Acoustic Guitar

Steel string acoustic guitars are closely related to classical guitars, but have slightly different structures. Some differences are stylistic and some are due to the higher tension required to bring steel strings to concert pitch. The neck usually crosses the body at the 14th fret and usually includes a truss rod, though not always an adjustable one.

The structure of the body also reflects the higher string forces; the top is generally thicker and the braces are larger. The most common bracing pattern for steel string acoustic guitars is called X-bracing. Figure 3.41 shows a guitar top with X-bracing. It also includes a proprietary (and patented) feature in the form of a channel routed around the edge of the top plate to increase flexibility near the glue joint with the sides.

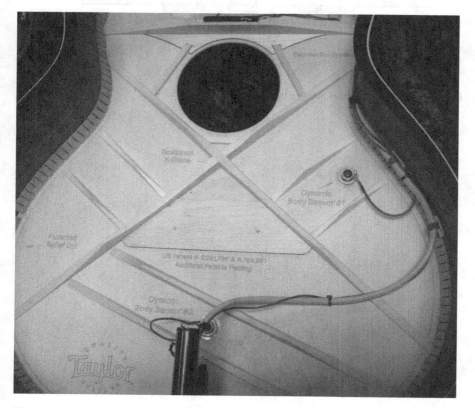

Figure 3.41 Steel String Guitar Top with X-Bracing (image by the author, reproduced courtesy of Taylor Guitars, www.taylorguitars.com)

The top in Figure 3.41 also shows the bridge plate. This distributes the load from the bridge and also forms a pad to help hold the strings on. Steel strings are terminated on the bridge end with a small metal bead and are known colloquially as ball-end strings. In most of the bridges on steel stringed acoustic guitars, the ball ends are pushed through vertical holes in the bridge and then held in place with tapered pins (called bridge pins) as shown in Figure 3.42.

The necks of steel string guitars have traditionally been attached to the body using a tapered dovetail joint as shown in Figure 3.43

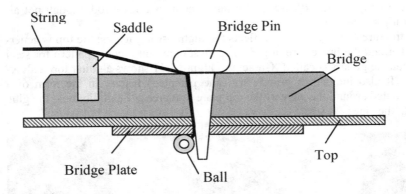

Figure 3.42 Ball End Steel String Fixed to Bridge

Figure 3.43 Dovetail Neck Joint (image courtesy Stewart-MacDonald, www.stewmac.com)

Figure 3.44 A Representative Bolt-on Neck (image courtesy Stewart-MacDonald, www.stewmac.com)

Increasingly, however, the necks of acoustic guitars are bolted to the body using recessed nuts in the heel. Figure 3.44 shows the parts of a representative bolt-on neck. In this design, there are bolts threaded into the tenon on the face of the heel and fit through holes drilled in the block.

An alternative design from Taylor Guitars (Figure 3.45) uses threaded inserts in the heel with bolts through the mounting block. In this design, the bolt is not permanently mounted to the neck; rather, it is inserted from the inside of the body through holes in the neck mounting block during final assembly. It also has an additional threaded insert in the lower portion of the fretboard extension of the neck that overhangs the body. Taylor uses this neck design along with a selection of precisely-manufactured spacers to ensure accurate and repeatable alignment of their instruments. The spacers, shown in Figure 3.46, are in integral part of the neck design so they drop in easily. They are produced in a range of thicknesses so that minor dimensional differences in the completed bodies and necks can be corrected precisely during final assembly.

Figure 3.45 Heel Showing Recessed Nuts for Bolted Neck Joint (image by the author, reproduced courtesy of Taylor Guitars, www.taylorguitars.com)

Figure 3.46 Neck Spacers at the Taylor Factory (image by author, reproduced courtesy of Taylor Guitars, www.taylorguitars.com)

3.1.3 Solid Body Electric Guitar

From a structural standpoint, the solid body electric guitar is very simple – the neck and body both have the primary job of supporting string tension and neither one needs to radiate acoustically. The neck is generally very similar to the type found on steel string acoustic guitars with the primary difference being that it is longer – on solid body guitars, the neck often crosses the body at the 16th fret. This offers players better access to the higher frets and offers the possibility of having more usable frets on the neck; some instruments have 24 frets – two full octaves.

While the sound is primarily electronic, it is not correct to say that the structure of the instrument has no effect at all. The choice of materials can color the sound and experienced players sometimes prefer specific materials.

One structural feature that sets solid body electric guitar apart from other types is the method of attaching the neck. The necks are either attached with long screws (called, in imprecise popular argot, a bolt on neck), glued to the body, or made as a single piece along with the body (called neck through body). Figure 3.47 shows a typical bolted-on neck with the screws running through a rectangular plate in the back of the body.

The method of connecting the neck to the body can affect sustain (the rate at which volume decreases). Conventional wisdom holds that the neck through body gives the longest sustain and the bolt-on neck gives the shortest. However,

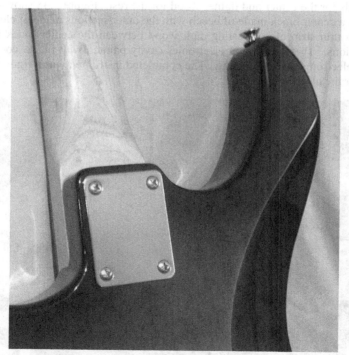

Figure 3.47   Neck-Body Joint of a Bolt-on Neck (Wikipedia Commons – image is in the public domain)

experiments by R.M. Mottola [56] suggest that a bolt-on neck offers the longest sustain of the three attachment methods. These results are interesting, but should be corroborated before being accepted as being generally applicable.

Sustain is essentially a measure of the structural damping in the instrument and it makes sense that some materials have more structural damping than others. Soft woods like alder and basswood (also called linden) are often used because they are light and make the instrument comfortable to play. However, a body made at least partially out of a harder wood like maple may slightly change the sound quality of the instrument and produce longer sustain. A well-known example is the different woods used for Fender Stratocasters.

Two popular species for Stratocaster bodies are alder and ash. Alder is a light, soft wood with a tight grain and guitars with alder bodies are sometimes described as having a rich or warm sound. Ash is a much harder wood with a more open grain. White ash is used for baseball bats and some types of ash (e.g. swamp ash) are widely used for electric guitar bodies. Guitars with swamp ash bodies are sometimes described as having a bright tone and good sustain.

Some solid guitar bodies use more than one type of wood. This is sometimes done for appearance, but can affect tone as well. A good example is the Gibson Les Paul in which the solid body is sometimes made of a relatively light wood and covered with a plate of maple or figured maple.

While it is not a common practice, some builders glue bodies together using hard woods for the center and lighter wood for the rest. Figure 3.48 shows a solid body with a center block made of beech with the outer sections made of clear pine. There is a thin strip of contrasting dark wood between the center block and the outer sections. There is also an electronics cavity painted with black, conductive paint for electromagnetic shielding. The completed instrument has a figured wood cover over the electronics pocket.

Figure 3.48 A Solid Body Made With both Hard and Soft Wood

3.1.4 Archtop Jazz Guitar

Archtop guitars are unique in that both the tops and backs are not flat, but rather convex like those of a violin [57]. They are popularly associated with jazz, but are also used for rock and popular music. Figure 3.49 shows a representative archtop instrument made by the author.

Archtop guitars can be thought of as an intermediate point between members of the violin family of instruments and conventional flat-top guitars. Both top and back plates are arched similar to those of a violin and are traditionally carved from solid wood blanks. Production instruments, such as the Gibson ES-335, generally use laminated top and back plates.

These instruments also generally have a different bracing pattern than flat-top instruments. Rather than a fan bracing pattern (classical) or X-brace (steel string), archtops typically use only two large braces – either side by side or in an X. They also do not use sound posts between the top and back as do members of the violin family. Figure 3.50 shows two bracing patterns suggested by Bob Benedetto [45]. While there other patterns are used, these can be considered as reasonably standard.

Traditionally, the tops and backs of archtop guitars have been carved from solid, bookmatched planks. In keeping with the tradition of violin makers, the tops are usually spruce and the backs are usually maple. Archtops are traditionally large instruments; bodies 21 in (533 mm) long and 17 in (432 mm) wide at the lower bout are common. The blanks from which the tops and backs are carved are typically 1–1 ¼ in (25.4–31.8 mm) thick.

Carving a top or back by hand can be a demanding process. About 80% of the material must be removed to result in a plate that is convex on the outside and concave on the inside. Furthermore, the thickness of the finished plate typically varies from something on the order of ¼ in (6.35 mm) thick at the center to

Figure 3.49 An Archtop Acoustic Guitar

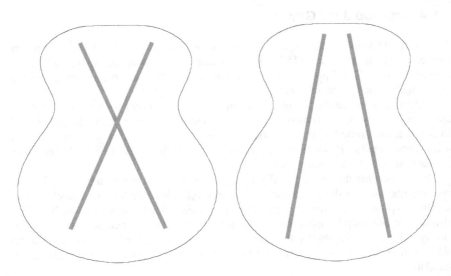

Figure 3.50 Typical Bracing Patterns for Archtop Guitars

about 1/8 in (3.18 mm) thick near the edge. Figure 3.51 shows a top being carved by hand and Figure 3.52 shows a cross section of an arched top (or back). At least one major manufacturer uses computer controlled mills to carve arched tops and backs from solid wood blanks.

Figure 3.51 An Arched Top Being Carved From a Solid Blank

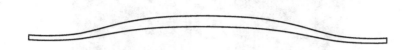

Figure 3.52 Typical Cross Section of an Arched Top or Back

Many mass production archtop guitars have tops and backs laminated from plies that have been pressed to the desired shape in heated molds, essentially like plywood. There is no reason that a fine guitar couldn't be made this way, but this method does produce plates with stiffness characteristics significantly different from those of carved plates. In a carved plate, the grain all runs the same direction while a laminated plate typically has cross plies. Also, the grain in the laminated plies is parallel to the surface of the plate while the grain in a carved plate isn't.

Another distinction among archtop guitars is that some are true hollow body instruments and some are not. For the example, the Gibson ES-335 (Figure 3.53), made famous by B.B. King among others, has a heavy wood block down the center of the body and the full depth of the body so it connects the top and the back. This was done to reduce feedback when playing a high volume and makes the instrument much more practical for rock musicians. Of course, the attendant increase in the stiffness of the body affects the sound, but the instrument is not meant to be played un-amplified. Note that the instrument in Figure 3.53 has a tailstop (block into which the ends of the strings are fixed) mounted directly to the soundboard. This would not be possible if there was not a significant internal structure.

True hollow body archtops are widely used by jazz musicians who favor a smooth sound. A hollow body archtop can't have heavy pickups mounted on the flexible soundboard, so they often have a pickup mounted on a raised pickguard rather than the body and positioned at the end of the neck. Figure 3.54 shows a Benedetto Manhattan archtop jazz guitar with a floating humbucker pickup.

Figure 3.53 Gibson ES-335 (Wikipedia Commons, image is in the public domain)

Finally, archtop guitars share a significant structural element with violins in that the strings are almost always attached to the instrument at the tail block using a tailpiece. As a result the force on the bridge is approximately normal to the top (Figure 3.55). Since the angles of the string on each side of the bridge are not generally identical, there is a small horizontal component of the resulting force. The bridge (called a floating bridge) is held against the top only by the string force; since it isn't glued to the top, there can be no moment applied to the top through the bridge. By not gluing the bridge to the top, it is possible to make large intonation adjustments by simply sliding the bridge back and forth. The obvious disadvantage is that the bridge might come loose while the instrument is being re-strung, requiring that the intonation be reset. This problem can be avoided by simply changing the strings one at a time. The instrument in Figure 3.54 has a floating bridge.

Figure 3.54 Benedetto Manhattan Archtop Jazz Guitar (image courtesy Benedetto Guitars, www.benedettoguitars.com)

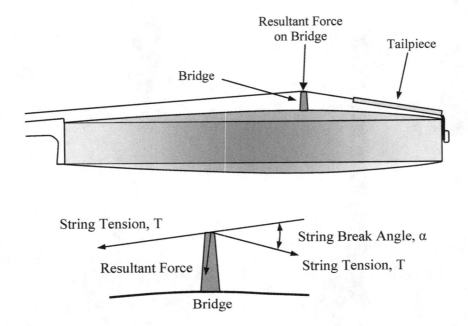

Figure 3.55 Force on Bridge of Archtop Guitar

3.1.5 Hybrid Instruments

At this writing, a class of acoustic-electric instruments is enjoying increasing popularity (see Chapter 1). This class of instruments is developing rapidly and their structural characteristics are far from standardized, so a general description is difficult. The bodies are generally thin to mimic the feel of a solid body electric guitar and are being made from both wood and composite materials. The necks of the most popular models are bolt on. Bracing patterns also vary; while some instruments use bracing patterns reminiscent of acoustic guitars, at least one uses no soundboard bracing at all.

Figure 3.56 shows a hollow body version of the Fender Stratocaster. This instrument is a development of Fender's iconic solid body instrument and uses a wood body with a graphite soundboard.

It is not uncommon for these instruments to have some combination of pickups installed in them and they often have on-board preamps. For example, the Taylor T5 has both soundboard sensors (essentially accelerometers) and magnetic pickups running through a specially designed preamp.

Figure 3.56   A Fender Acoustasonic™ Strat® (image courtesy of Fender Musical Instrument Corp., www.fender.com)

3.2 Static and Dynamic Loads

The loads on a guitar are dominated by string forces. The magnitude of the string forces depends on the length and diameter of the strings, their composition, their pitch and how many strings are used. Different string diameters can give very different tonal qualities and players are often very particular about the size and type of strings they use. Strings are generally supplied in matched sets and large manufacturers may stock scores of different string sets. In the vernacular, sets are often identified by the size of the lightest string; it is common to hear someone say that they like 'playing nines', referring to string sets whose lightest string has a diameter of 0.009 in (0.23 mm). Larger diameter strings require higher tension to bring them to standard pitch. Table 3.3 gives representative tensions for a range of strings.

Table 3.3 Tension From Representative String Sets [42]

Type	E lb/kg	B lb/kg	G lb/kg	D lb/kg	A lb/kg	E lb/kg	Total lb/kg
XLS530 Extra-Super Light (Electric)	10.4/ 4.72	9.1/ 4.13	12.9/ 5.85	11.9/ 5.40	13.9/ 6.30	11.9/ 5.40	70.1/ 31.8
EXL110 Regular Light (Electric)	16.2/ 7.35	15.4/ 6.98	16.6/ 7.53	18.4/ 8.34	19.5/ 8.84	17.5/ 7.94	103.6/4 7.0
EJ22 Jazz Medium (electric)	27.4/ 12.4	26.3/ 12.0	32.8/ 14.9	34.8/ 15.8	31.1/ 14.1	26.3/ 11.9	178.7/8 1.1
EJ11 Light (Steel/Bronze Acoustic)	23.3/ 10.6	23.3/ 10.6	29.4/ 13.3	29.5/ 13.4	28.4/ 12.9	25.1/ 11.4	159.0/7 2.2
EJ18 Heavy (Steel/Bronze Acoustic)	31.8/ 14.4	29.5/ 13.4	38.4/ 17.4	45.2/ 20.5	40.0/ 18.1	32.2/ 14.6	217.1/9 8.4
EJ43 Light (Nylon/Silver Classical)	14.8/ 6.71	11.2/ 5.08	11.7/ 5.31	14.8/ 6.71	12.5/ 5.67	13.2/ 5.99	78.2/ 35.5
EXP44 Extra Hard (Composite/Silver)	16.4/ 7.44	12.5/ 5.67	12.9/ 5.85	16.8/ 7.62	16.6/ 7.53	16.8/ 7.62	92.0/ 41.7

While string tension is the most obvious static load on the instrument, it is not generally the only one. Almost all engineering materials expand or contract with temperature changes and this behavior is generally expressed in terms of a coefficient of thermal expansion [58]. For a long, thin rod, the change in length due to a change in temperature is expressed as

$$\Delta L = \alpha \, L \, \Delta t \tag{3.21}$$

Where L is the length, Δt is the change in temperature and α is the coefficient of thermal expansion. For mild steel, $\alpha = 6.5 \times 10^{-6}$ in/in/°F (11.7×10^{-6} mm/mm/°C). The coefficient of thermal expansion for wood (parallel to the grain) is largely independent of species and density [43] and ranges from 1.7×10^{-6} in/in/°F to 2.5×10^{-6} in/in/°F (3.06×10^{-6} mm/mm/°C to 4.50×10^{-6} mm/mm/°C).

The expression for change in length as a function of applied force can be substituted to form a relation between change in force and change in length.

$$\Delta L = \frac{\Delta F L}{AE} \quad \Rightarrow \quad \Delta F = \frac{\Delta L A E}{L} = \alpha \Delta T A E \tag{3.22}$$

This difference in expansion coefficients is not generally a problem, though there can be exceptional cases of large temperature changes such as when guitars are checked into the baggage hold on airliners. Since air at cruising altitude is very cold, it is possible to have a temperature differential of 60°F (33°C). Note that this is a temperature differential, not temperature; 60°F on a thermometer corresponds to 15.6°C, but a temperature differential in °C is simply 5/9 of the differential in °F. For a typical wood guitar, this would induce an increase in tension of 2 1 b–10 l b (8.9 N–44.5 N) in each string.

The increase in tension is a strong function of cross sectional area, so strings with a larger load carrying area (windings don't count, only the core) generate larger forces as temperature decreases. If an instrument has a existing damage or weak glue joints, the increased tension can cause a structural failure. Indeed, most manufacturers advise that strings be detuned (tension decreased) before checking instruments into an airliner baggage hold. Note that the instrument will probably have warmed significantly by the time it has been retrieved. If so, the string tensions and the resulting frequencies will have returned to normal.

Another source of structure failure is changes in humidity. The ideal building environment is approximately 70°F (21°C) and 45% relative humidity (RH). If an instrument built at 45% RH is exposed to very low humidity, the wood dries and tries to contract. However, since the instrument is generally made of many components glued together, not all parts can contract as they otherwise would and internal stresses are created. A common failure is splitting of the back or soundboard. As an anecdotal piece of evidence, the author currently works in a building where the RH occasionally drops below 20% in the winter. During one winter, three of the instruments in his office suffered catastrophic splits in the soundboards.

The dynamic loads from the vibrating strings are harder to quantify than the static loads since they vary with string tension, mass and playing style. In qualitative terms, dynamic forces are generally much smaller than static forces. Unless an instrument is being played in an extremely enthusiastic manner (think Pete Townshend), it is extremely unlikely that dynamic forces could cause structural problems. Of course, the dynamic forces are the ones that produce sound and are, thus, quite interesting.

To produce sample data, the dynamic forces perpendicular to the string were measured using a simple test fixture and a dynamic force sensor [59] as shown in Figure 3.57.

The results for two different types of strings are presented in Figures 3.58a and 3.58b. Content below 40 Hz has been removed from both signals so that the effect of string vibration can be clearly seen. Figure 3.58a shows the time domain force response of a wound A string for an electric guitar after being plucked at the center of the string. Part of the initial pluck was attenuated by the high pass filter, but the maximum force applied was on the order of a few Newtons. However, the response force level never exceeds approximately 0.7 N (0.16 lb). Figure 3.58b shows the time domain force response of a wound nylon low E string, also plucked at the center. Though the shape of the envelope differs, the magnitude of the force is roughly equal.

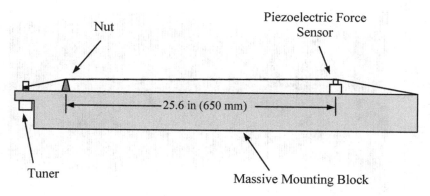

Figure 3.57 Test Setup for Measuring Dynamic String Force

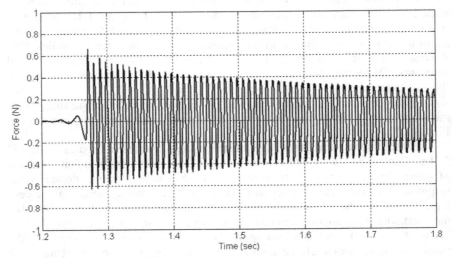

Figure 3.58a Dynamic Normal Force on Bridge from Steel A String (40 Hz Highpass)

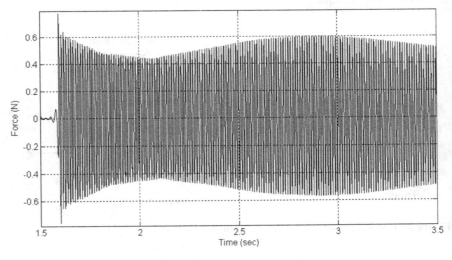

Figure 3.58b Normal Force on Bridge from Nylon Low E String (40 Hz Highpass)

It is important to note that a plucked string generally does not vibrate only in one plane. The force gauge is only sensitive in one direction and was installed in the fixture to measure the vertical component of the dynamic force – the component normal to the soundboard. The test was set up this way because the bridge and the soundboard are extremely stiff in the lateral direction and can be assumed to move only out of plane. The analysis that accompanies this experiment is not very involved and appears in the following chapter.

3.3 Materials

Traditionally, all guitars have been made of wood. Unlike most violin makers, guitar makers have been quite willing to make instruments from just about any wood available. For acoustic guitars, the tops are generally made from either spruce or cedar. Traditionally, backs and sides have been made from either rosewood, maple or mahogany and it is common on individual guitars for the back and sides to be made of the same material.

Necks need to withstand compressive and bending loads while not having an adverse effect on the weight and balance of the instrument. Mahogany and maple are the most commonly used materials. It is important to note that there are several species of maple in common use; hard maple is used for necks while soft maple is typically used for backs and sides. The figured maple used for violin backs, faces of solid body guitars and backs and sides of both flat top acoustic and archtop guitars is soft maple.

Electric guitar bodies have much less effect on the sound of the instrument, so the choice of materials is dictated more by weight and appearance. The list of materials that have been used is far too long to catalog here, but alder and swamp ash are popular. It is also common, for the sake of aesthetic appeal, to overlay the body with a thin piece of some figured wood. Figured maple (Figure 3.59) is perhaps the most common choice.

As the costs of traditional materials increase, luthiers have been enthusiastic about seeking new alternatives. A wide range of tropical hardwoods is routinely being used for the backs and sides of acoustic instruments and for the fretboards and body overlays on electric guitars. Encouragingly, there is an international treaty in place (Convention on International Trade in Endangered Species, CITES) [60] that seeks to regulate trade in (among a large very group of flora and fauna) hardwood used for musical instruments. Tropical wood used for guitars should be harvested and sold in accordance with CITES. While this should ensure long term supplies of wood, it does limit the current supply; prices are accordingly high and will likely remain so unless some alternative material is accepted by the market.

Figure 3.59 Fender Showmaster Guitar with a Highly Figured Maple Top (image courtesy Fender Musical Instrument Corp., www.fender.com)

Figure 3.60   A Large Sitka Spruce Log Being Salvaged for Instrument Wood (image courtesy Brent Cole Sr., Alaska Specialty Woods, www.alaskawoods.com)

Perhaps the most important facet of using high quality wood to make musical instruments is ensuring reliable supplies. At this writing, nearly 1 million new instruments are sold in the United States each year, with the associated requirements for raw materials. Softwoods are widely farmed and harvested sustainably for use as lumber. However this wood is not high enough quality to be used for guitars.

The soft woods used for acoustic guitar tops have traditionally been cut from old-growth trees, usually from the Pacific Northwest. The ideal top is quartersawn from a large, clear, straight log with closely-spaced growth rings. Two preferred species are Sitka spruce and western red cedar. Logs of high enough quality to yield instrument grade wood are not common since knots, cracks or pitch pockets are unacceptable. The best wood often comes from trees several hundred years old. Trees this old are now rare and often protected from logging

Happily, there are a number of suppliers cutting wood for tops from salvaged logs (Figure 3.60). Sitka spruce was used for routine structural applications such as bridges and salmon traps, and large logs can be recovered and processed for use in musical instruments.

4 Dynamic Behavior

The sound of a guitar is intimately related to the dynamic response of its structure and its interaction with the air both in and around it. The problem can be simplified, at least initially, by considering the dynamics of the strings and the body separately. This chapter introduces some basic ideas in structural dynamics, describes the dynamic behavior of strings and bodies and then briefly summarizes the response of the complete instrument.

4.1 Structural Dynamics

Understanding the dynamic response of the structure is at the core of understanding how the instrument produces sound. One of the most fundamental concepts is that of natural frequencies and mode shapes. Any elastic structure has frequencies at which it 'likes' to vibrate. More precisely, these are frequencies at which a low level of input force results in a large response amplitude and are variously called natural frequencies, resonant frequencies or eigenvalues [61]. The last term is because resonant frequencies are solutions to the eigenvalue equation that results from structural models of the instrument. The eigenvalue equation has the form $Kx = \omega^2 Mx$, where K describes the stiffness of the structure, M describes its mass, x describes the displacement and ω is the resonant frequency. Math models that lead to the eigenvalue equation are discussed in the next chapter.

By applying a known input force to the structure and measuring the response, it is a simple thing to produce a frequency response function (FRF) that clearly shows the resonant frequencies. Figure 4.1 shows an FRF measured from a Taylor Model 414CE. The input was provided by an instrumented hammer tapping at the right side of the bridge. The response was measured by a laser displacement sensor observing a point right next to the impact point near the right side of the bridge. Note that the impedance (dynamic stiffness) of the top varies with position. If the test position was somewhere else on the top, the frequencies of the peaks would be unchanged, but the shape of the FRF would be different. Figure 4.2 shows the test setup.

R.M. French, *Engineering the Guitar*,
DOI: 10.1007/978-0-387-74369-1_4, © Springer Science+Business Media, LLC 2009

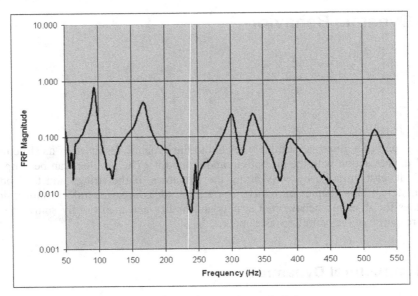

Figure 4.1 Frequency Response Function from an Acoustic Guitar

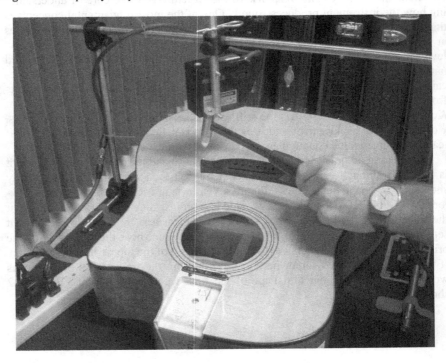

Figure 4.2 Measuring the FRF of a Guitar Body before the Neck is Mounted (photo the author, reproduced courtesy of Taylor Guitars, www.taylorguitars.com)

The peaks at approximately 95 Hz, 170 Hz, etc. are the resonant frequencies of the instrument. The peaks indicate frequencies for which the motion of the structure per unit input is large; these are frequencies at which the structure 'likes' to vibrate [62]. A more complete description of the lower resonant frequencies of an instrument appear later in this chapter.

An FRF can be calculated when a structure is made to vibrate (excited) in such a way that both the input force and the resulting motion (response) are recorded. The FRF in Figure 4.1 was calculated from data recorded when a guitar was tapped with an instrumented hammer and the response was recorded using a laser displacement sensor. The input and output data are initially in the form of time-varying voltages, often referred to in technical literature as signals.

The FRF is, by definition, a function of frequency, but all data is, by definition, recorded in time domain. After all, we live in the time domain and it seems quite unlikely that any of us knows how to live in the frequency domain. Rather, time domain data is transformed into the frequency domain using a Fast Fourier Transform (FFT) [63]. This is a standard feature of any signal processing software and is so ubiquitous that even Microsoft Excel has an FFT function.

Let us call the time domain input signal x and the time domain output signal y. Capital letters are often used to represent signals that have been transformed into the frequency domain, so let X be the input signal after it has been transformed into the frequency domain and let Y be the output signal in the frequency domain. The FRF is generally represented by the letter H and is simply the ratio of output to input in the frequency domain.

$$H(\omega) = \frac{Y(\omega)}{X(\omega)} \tag{4.1}$$

Equation (4.1) is the simplest expression of the FRF [64]. To minimize the effects of unwanted noise in the measurement system, more sophisticated ways of calculating H are often used. The details are not important here, but the two most common forms of FRF are the H_1 and H_2 forms. These are intended to minimize the effect of noise on the output and input respectively and are, for the purposes of this discussion, equivalent to the expression in Equation (4.1).

The resonant frequencies offer insight into the dynamic response of an instrument, but there is another important component. Corresponding to each resonant frequency, there is a unique shape in which the structure vibrates. This is called either a mode shape or an eigenvector. Physically, the mode shape is the vibrational shape that results when the structure is excited at the corresponding resonant frequency. If a structure vibrating at a resonant frequency were illuminated with a strobe light pulsing at the same frequency so that the motion appeared to be frozen at its highest amplitude, that frozen shape would be the mode shape.

The string dynamics are the primary dynamic structural characteristic of electric guitars since the body is generally stiff and massive in comparison. For acoustic guitars, though, the dynamic response of the body and the body's coupling with the air are critical components of sound production.

4.2 Strings

The vibration of strings is perhaps the most fundamental element in understanding the behavior of the instrument as a whole. Vibrating strings have been used to make musical instruments throughout the recorded history of man [65].

Vibrating strings are particularly suited to making music because of their dynamics; a string is the only simple structure whose resonant frequencies are integer multiples of the fundamental. The ideal string is uniform along its length and has no bending stiffness, so the only stiffness is provided by the tension. The transverse motion of the string is assumed to be small when compared to its length and the tension is assumed to be constant. With these assumptions, the natural frequencies of stretched string are

$$\omega_n = \frac{n\pi}{L}\sqrt{\frac{T}{\rho}} \quad and \quad f_n = \frac{n}{2L}\sqrt{\frac{T}{\rho}} \tag{4.2}$$

Where L is the length of the string, n is an integer greater than zero, T is tension and ρ is mass per unit length (*not* per unit volume). This expression is the result of solving a simple differential equation and necessarily doesn't predate the development of calculus. However, the basic relationships between string frequency, length and tension were described by Galileo decades before the birth of Isaac Newton in 1642 [66].

Equation (4.1) defines the static loads on a guitar. Consider an E-string made of steel wire, d = 0.010 in (0.254 mm) and L=25.5 in (648 mm). At standard pitch, the fundamental frequency of the string is tuned to E_4, 329.6 Hz. The expression for tension comes from simply re-arranging Equation (4.2)

$$T = 4\rho L^2 f^2 \tag{4.3}$$

If one assumes a nominal value for the density of steel, then T = 16.3 lb (72.6 N). The diameters of the strings vary and the lower pitched strings are wound with wire to increase ρ. The result is that the tension in each of the six strings is approximately equal. Some string manufacturers helpfully list tensions on the packages. A typical example is a package of regular light acoustic strings that lists tensions ranging from 15.4 lb (68.5 N) to 19.2 lb (85.4 N). The total tension from all six strings in this set is 102.8 lb (457.3 N).

Since the string is assumed to have no bending stiffness, its ends always have pinned boundary conditions. Happily, the mathematical description of the mode shapes, called here *y(x)*, is nearly as simple as that for the natural frequencies Equation (4.4). Figure 4.3 shows the first five mode shapes.

$$y(x) = \sin(n\pi x / L) \tag{4.4}$$

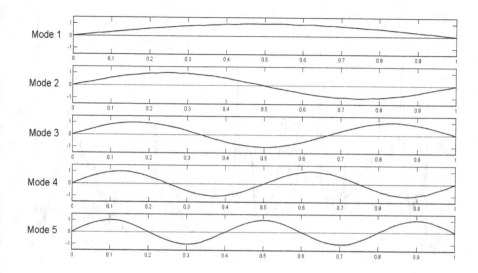

Figure 4.3 First Five Mode Shapes of an Ideal String

Note that even-numbered modes have a node line at the center of the string. Thus, it's very difficult (theoretically impossible) to excite those modes by pluck-ing the string at its midpoint – the 12th fret. Indeed, all but the first mode has a node line somewhere along the string. This suggests something that guitarists know instinctively, that changing the point where the string is plucked changes the resulting tone. Similarly, changing the location of the pickup also changes the tone. The natural frequencies of the vibrating string are present in different pro-portions when observed at different positions. This is why electric guitars are of-ten fitted with several pickups and a switch so the player can choose among them.

Figure 4.4 shows the frequency content of the measured instrument response due to a string plucked at the 12th fret and the 18th fret. The response was meas-ured using an accelerometer on the soundboard, so it also includes the dynamics of the instrument itself. Also, the response was normalized so that the maximum amplitude of each time domain signal is 1. The data is from a D string with a re-sonant frequency of 146.8 Hz and the harmonics are clear at 293.7 Hz, 440.5 Hz, etc. The shapes of the two curves in Figure 4.4 are complex, reflecting the nature of the physics that produced them. For now, the important feature is that the am-plitudes of the different curves are different at most frequencies. Since the in-strument and the test method were unchanged between the two recordings, the dif-ference in response is due to the difference in the plucking location.

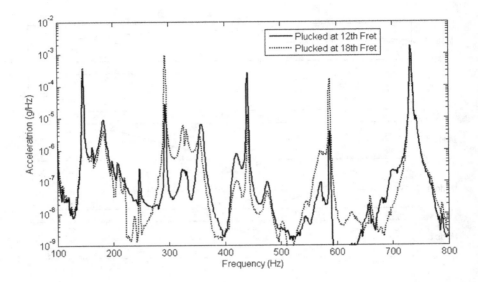

Figure 4.4 Effect of Plucking String at Different Locations

4.2.1 Analyzing String Dynamics

Equation (4.2) is very useful in describing the behavior of vibrating strings and of the forces acting on a guitar, so it makes sense at this point to understand where it came from. Fortunately, it is reasonably simple to solve the governing equation for an ideal string and estimate both the static and the dynamic forces applied to the bridge. The description here follows that presented by Fletcher and Rossing [25]. The string motion is described by the one-dimensional wave equation.

$$\frac{\partial^2 y}{\partial t^2} = \frac{T}{\rho} \frac{\partial^2 y}{\partial x^2} = c^2 \frac{\partial^2 y}{\partial x^2} \tag{4.5}$$

Where T is string tension, ρ is the mass per unit length and c is the speed of wave propagation. Additionally, y is the displacement of the string and x is the position along the string.

Equation (4.5) is a partial differential equation, the general group of which is notorious for intimidating science and engineering students. However, this one is particularly easy to solve. Solving the equation means finding a function of position and time, $y(x,t)$, that satisfies the equation. Once this function is identified, it can be used to study the motion of the string and the forces it applies to its end supports – the nut and saddle.

It is important to note at this point that there are two fundamentally different kinds of waves at work when a guitar is played; physicists call them dispersive and non-dispersive waves. A dispersive wave is one in which the speed of propagation is a function of the frequency. For example, bending waves in a structure (like a guitar top) and ocean waves are dispersive. Non-dispersive waves are ones in which the speed of propagation is uniform no matter the frequency. Sound traveling through air is non-dispersive. Since the wave speed, c, in Equation (4.5) is a constant, it is clear that a wave propagating through a stretched string is also non-dispersive.

Several solution methods for the 1-D wave equation are generally reported in books on applied math [61] and two of them are worth mentioning here; one is called separation of variables and the other is called D'Alembert's method after the 18th century French mathematician Jean le Rond D'Alembert, who proposed it.

The wave equation belongs to a category of differential equations called separable equations. Separable equations have solutions of the form

$$y(x,t) = F(x)G(t) \tag{4.6}$$

Where $F(x)$ and $G(t)$ are different functions, each depending on only one variable. When this general solution is substituted into the 1-D wave equation, the result is

$$F\frac{d^2G}{dt^2} = c^2 G\frac{d^2 F}{dx^2} \tag{4.7}$$

Which is easy to rewrite as

$$\frac{1}{c^2 G}\frac{d^2 G}{dt^2} = \frac{1}{F}\frac{d^2 F}{dx^2} \tag{4.8}$$

The left side of this equation is a function only of time and the right side is a function only of position. Because of this, both sides must be constant. This part of the argument tends to sneak up on you, but it is correct. If the terms weren't constant, then changing t would affect the left side of the equation, but not the right side; they would then not be equal to one another. Similarly, changing x would only affect the right side of the equation, and with the same inadmissible result. Thus, it is correct to set both sides of Equation (4.8) equal to some constant. If that constant is called k, then the original partial differential equation can be written out as two uncoupled ordinary differential equations

$$\frac{d^2 F}{dx^2} - kF = 0$$

$$\frac{d^2 G}{dt^2} - c^2 kG = 0 \tag{4.9}$$

Since the displacement, y, is a function of both time and position, the 1-D wave equation cannot be solved unless there is some specific information about the string in both space and time. These are its boundary conditions and initial conditions.

The nut and saddle form the endpoints of the string. Thus, for any time, $y(0,t)$ = $F(0)$ = 0 and $y(L,0)$ = $F(y)$ = 0, where L is the length of the string. As the string is plucked, it is deformed to some shape we will call $u(x)$ and released. In the instant before the string is released, its velocity is zero. Thus at time $t = 0$, the initial conditions are $y(x,0)$ = $G(0)$ = $u(x)$ and $v(x,0)$ = $\dot{G}(0)$ = 0, where $v(x,t)$ is the transverse velocity of the string and $\dot{G}$ is the time derivative of G.

Boundary conditions are applied to the first equation and initial conditions are applied to the second equation. After some manipulation too detailed to reproduce here, the solution can be shown to be

$$y(x,t) = \sum_{n=1}^{\infty} B_n \cos\left[\frac{cn\pi t}{L}\right] \sin\left[\frac{n\pi x}{L}\right]$$

$$= \frac{1}{2}\sum_{n=1}^{\infty} B_n \sin\left[\frac{n\pi}{L}(x-ct)\right] + \frac{1}{2}\sum_{n=1}^{\infty} B_n \sin\left[\frac{n\pi}{L}(x+ct)\right]$$

(4.10)

The second form of the solution presented in Equation (4.10) is the result of applying a basic trigonometric identity. Note that, in the upper expression above, $\cos(cn\pi t/L)$ = $\cos(\omega t)$, so ω=$cn\pi/L$, in agreement with Equation (4.2). Similarly, $\sin(n\pi x/L)$ in the same expression corresponds to Equation (4.4). The constant B_n is the coefficient in a Fourier sine series and is defined as

$$B_n = \frac{2}{L}\int_0^L u(x)\sin\left(\frac{n\pi x}{L}\right)dx$$

(4.11)

There is another means of arriving at this analytical solution to the 1-D wave equation. In the 18th century, Jean le Rond D'Alembert showed that the general solution, assuming zero initial velocity, is

$$y(x,t) = \frac{1}{2}\left[f_1(ct - x) + f_2(ct + x)\right]$$

(4.12)

Readers interested in more detail are encouraged to review the development presented by Kreysig [61]. Clearly, Equations (4.10) and (4.12) are equivalent if

$$f_1 = \sum_{n=1}^{\infty} B_n \sin\left[\frac{n\pi}{L}(x-ct)\right]$$

$$f_2 = \sum_{n=1}^{\infty} B_n \sin\left[\frac{n\pi}{L}(x+ct)\right]$$

(4.13)

To a first approximation, the force exerted by a vibrating string on its end supports is dues to the slope at the end supports [25]. If the angle of the string at the support is θ and the angle is small, the transverse force is

$$F_T = T\sin\theta \qquad (4.14)$$

And the longitudinal force is

$$F_L = T\cos\theta \qquad (4.15)$$

These expressions also assume the string tension is constant (as does the original 1-D wave equation). This is not strictly true, but is close enough for our purposes here. The next step is to use these expressions to analyze a representative string.

Assume a steel string with a length of 648 mm (25.5 in) and a diameter of 0.41 mm (0.016 in) and a tension of 104N (23.3 lb). Assume also that the string is displaced 3 mm (0.12 in) at a point $L/4$ from the saddle as shown in Figure 4.5.

The series expression for the initial displacement of the string is

$$u(x) = \sum_{n=1}^{\infty} B_n \sin\left[\frac{n\pi x}{L}\right]$$
$$= \left[\frac{32}{n^2\pi^2}\sin\left(\frac{3n\pi}{4}\right) - \frac{24}{n^2\pi^2}\sin(\pi n)\right]\sin\left[\frac{n\pi x}{L}\right] \qquad (4.16)$$

The displacement calculated from this expression is presented in Figure 4.6 at the instant the string is plucked and at 100μsec increments afterwards.

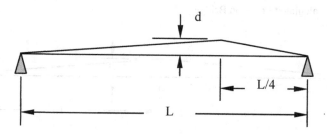

Figure 4.5 Initial Displacement of a Plucked String

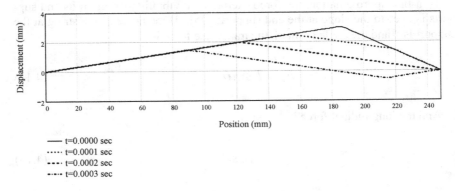

— t=0.0000 sec
····· t=0.0001 sec
---· t=0.0002 sec
-···· t=0.0003 sec

Figure 4.6 Calculated Deformation of a Plucked String as a Function of Time

Figure 4.7 shows the calculated transverse force on the bridge. The magnitude is higher than the measured values shown earlier, but still serves to show that the transverse force is small compared to the string tension – about 5% in this case. A more sophisticated analysis (variable tension, bending stiffness, etc.) would presumably correspond more closely to experiment.

If the calculations are repeated for a string plucked at the 12th fret (the mid point of the string), the initial condition is as shown in Figure 4.8.

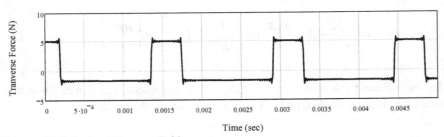

Figure 4.7 Calculated Force on Bridge

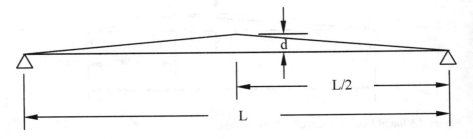

Figure 4.8 Initial Condition of String Plucked at Center

The resulting force at the bridge as a function of time is shown in Figure 4.9.

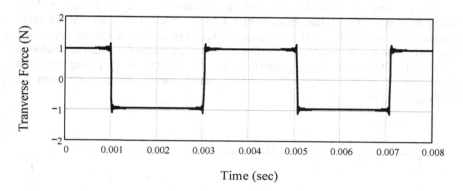

Figure 4.9 Calculated Force on Bridge of a String Plucked at the Center

Figure 4.3 shows that string modes 2 and 4 have node lines at the center. In fact, all the even numbered modes had nodal points at their centers. It follows then that it should not be possible to excite these modes by plucking a string at its mid point. This is confirmed by the Fourier coefficients (see Equation (4.11)), B_n. B_2, B_4 and so on are all zero as shown in Figure 4.10.

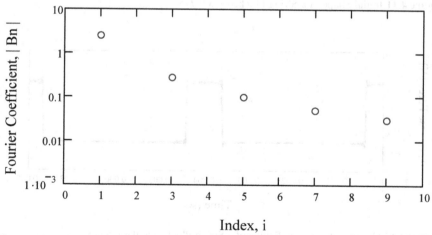

Figure 4.10 Fourier Coefficients for a String Plucked at the Center Showing Missing Even-Numbered Terms

The calculation is repeated for a string plucked at a position $L/5$ from the bridge. The initial shape is shown in Figure 4.11 and the resulting calculated force is shown in Figure 4.12.

It is perhaps not surprising that the fifth mode is not present in the resulting string motion. This is clear from Figure 4.13 in which B_5 is zero. In the same manner that none of the even numbered modes were present when the string was plucked in the center, modes 5, 10, 15, etc. are absent when the string is plucked a $x = L/5$. By extension, plucking the string at $x = L/n$ (where n is an integer) results in $B_n = B_{2n} = \ldots = 0$. Figure 4.13 shows that B_5 and B_{10} are missing from the string plucked at $x = L/5$.

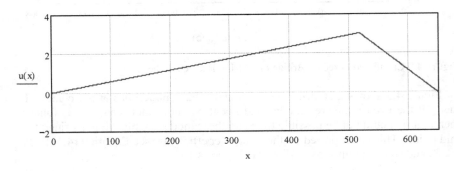

Figure 4.11 Initial Shape of a String Plucked at $x = L/5$

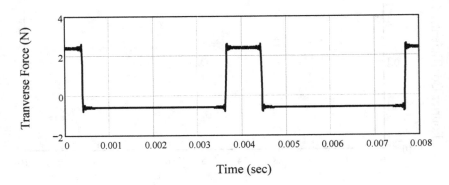

Time (sec)

Figure 4.12 Calculated Force on Bridge of a String Plucked at the $x = L/5$

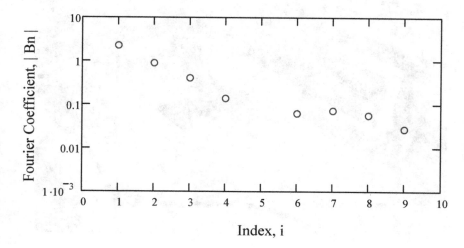

Figure 4.13 Fourier Coefficients for a String Plucked at $L/5$

In practical terms, this means that changing the point at which the string is plucked changes the frequency content of the vibrating string and, thus, the resulting tone. It is easy to confirm that plucking a string at the 12th fret (the center of the string) produces as smooth tone. Conversely, plucking the string near the bridge produces a brighter tone with more high frequency content.

In the same way that plucking a string in different places affects the frequency content, the placement of pickups on an electric guitar affects which string modes can be detected. A pickup placed near the neck produces a smoother tone with less high frequency content while one placed near the bridge produces a bright, clear tone. It is common for electric guitars to have two or three pickups and a multi-position switch. The player can select different combinations of pickups in order to have more control over the tone. At the other end of the scale, some archtop jazz guitars have a single pickup mounted to the end of the neck (see the instrument in Figure 3.54). This is intended to produce the smooth tone favored by many jazz players.

4.2.2 Experimental Results from Vibrating String

With the results of the analysis of the ideal string in hand, it is helpful to compare them to experimental data from a real string tested under ideal conditions. The dynamics of a string on a guitar reflect the fact that the string is only part of a

Figure 4.14 Test Setup for Idealized String

vibrating system that includes the guitar structure and the air both in and around it. The string behavior is much closer to the ideal when the string is mounted on a fixture designed to provide rigid, massive end conditions. Figure 4.14 shows a test in which steel string is mounted on a massive block and supported on each end by heavy steel wedges. The block is also glued to the bench top to increase its mass and stiffness.

The string is very light and flexible compared to the structure that supports it, so it is a reasonable approximation to the idealized string. It is important that no mass be added to the string during the test and that no external forces are applied. Thus, conventional sensors (accelerometers, inductive pickups, etc.) are not acceptable. Rather, the string motion was observed using a two channel laser Doppler vibrometer that uses the frequency shift of reflected laser light to measure the velocity of the target. One channel observed the horizontal motion of the string and the other observes the vertical motion. The measurement points on the string are shown in Figure 4.15.

Figure 4.15 Measurement Points on the String

The string was tuned to 329.6 Hz (E$_4$) and plucked near one end in order to ex-cite a large number of modes. Figure 4.16 shows the time domain response of the string. The purpose of the heavy mounting fixture was to minimize the energy lost to the fixture as the string vibrated and this is reflected in the very low damp-ing shown in the measured response. Low damping is easily observed as in-creased decay time – a note being audible for an extended time after being played.

The string exhibits approximately exponential decay, as one would expect from a proportionally damped structure. The envelope also suggests that two frequen-cies are generating a slow beat phenomenon [62].

Figure 4.17 shows a spectrogram of the same data set. The fundamental mode of the string is visible for about 16 seconds – a very long decay time and much more than one would see from a string mounted on an acoustic guitar. Addition-ally, the decay time is generally longer for steel strings than for nylon strings since nylon has a higher material damping.

In addition to the fundamental frequency, harmonics up through almost the limit of the human hearing range were observed (for clarity, only frequencies below 10 kHz are presented here). Not all harmonics have been excited and their amplitudes vary. It is exactly these variations that give a note its characteristic timbre.

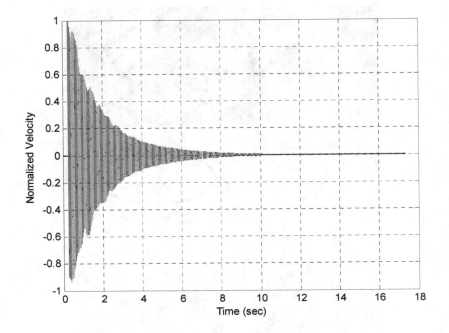

Figure 4.16 Velocity Response of Plucked String

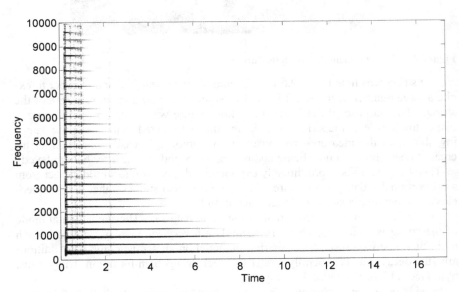

Figure 4.17 Spectrogram of String Response

While the resonant frequencies displayed in Figure 4.17 appear to be equally spaced, as predicted by Equation (4.2), there is a slight variation with increasing frequency. If the string behaved exactly as the ideal string approximation, every frequency would be an integer multiple of the fundamental. However higher harmonics are increasingly sharp as shown in Figure 4.18.

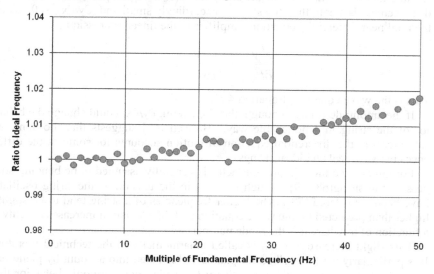

Figure 4.18 Measured Inharmonicity of a String

The assumptions leading to the expression for resonant frequencies of an ideal string are:

- Zero bending stiffness
- Small displacements
- Constant tension

Differences between the predicted and actual frequencies are assumed to arise from the fact that these assumptions are not exactly satisfied.

The differential equation describing the motion of a vibrating string [67] with no bending stiffness but with no limitations on displacement is

$$\frac{\partial^2 y}{\partial t^2} = \left[c^2 + c_0^2 \left[\sqrt{1 + \left(\frac{\partial y}{\partial x} \right)^2} - 1 \right] \right] \frac{\partial^2 y}{\partial x^2} \qquad (4.17)$$

Where

$$c = \sqrt{\frac{TL}{m}} \qquad and \qquad c_0 = \sqrt{\frac{E}{\rho}} \qquad\qquad (4.18)$$

Note that m is total mass of the string and ρ is the mass per unit length. If the displacement is small, then the slope is accordingly small and $(\partial y/\partial x)^2 \approx 0$. Then, the nonlinear differential equation simplifies to the linear expression

$$\frac{\partial^2 y}{\partial t^2} = c^2 \frac{\partial^2 y}{\partial x^2} \qquad\qquad (4.19)$$

This is the wave equation (Equation 4.5).

If the amplitude is large enough, the slope term, $\partial y/\partial x$, could change the behavior of the string in a noticeable way. Bolwell [67] suggests that the nonlinear slope term in the differential equation of motion accounts for some of the differences between real and ideal strings.

For guitars, the more important factor is generally assumed to be bending stiffness of the string[68, 69], though changes in the tension as the string oscillates have been considered [70]. The higher frequencies of a string tend to be slightly higher than predicted by the wave equation and this deviation increases roughly as a function of n^2 where n is the mode number.

This slight increase in pitch is called inharmonicity in the technical literature. It is particularly well-described for pianos and is taken into account by piano tuners. Short strings are much more likely to display non-harmonic behavior than long ones, so tuning a piano is necessarily an exercise in minimizing the errors in the most pleasing way. A well-tuned piano typically matches an even tempered scale in the range from C_4 (middle C, 261.6Hz) to A_4 (440 Hz). The higher notes are slightly sharp and lower ones are slightly flat. The approximate relationship between the notes and inharmonicity in the piano is called the Railsback curve after O.L. Railsback, author of the paper in which it first appeared in 1938 [71].

Short strings display more inharmonicity than do long ones of the same diameter. It is the short treble strings in the piano that are sharpest. It is then no surprise that the most prized pianos are the one with the longest strings. Concert grand pianos can be as much as 10 feet long (305 cm). Of course, there are any number of practical inconveniences associated with a piano the size of a small car, so upright and even smaller spinet models are popular. However, their tone is almost universally inferior to large concert grands. When was the last time a concert pianist gave a public performance on a spinet piano?

It is worth noting that inharmonicity is usually defined in cents where one cent is 1/100 of an even tempered half-tone. This unit is universally used by musicians and electronic tuners quite often display pitch in cents sharp or flat (see Figure 4.19).

Figure 4.19 Electronic Tuner with ±40 Cent Scale

The ratio of actual to ideal frequencies can be expressed as

$$\frac{f_n}{nf_0} = 2^{\delta/1200} = e^{\delta/1731} \approx 1 + \delta/1731 \qquad (4.20)$$

And the deviation, δ, expressed in cents is

$$\delta = 1731 \ln(f_n / nf_0) \approx 1731(f_n / nf_0 - 1) \qquad (4.21)$$

Assuming constant tension, but finite stiffness, the approximate deviation can be cast in a closed form expression

$$\delta = \frac{1731}{2} \frac{\pi^2 E \kappa^2 A n^2}{L^2 T} = bn^2 \qquad (4.22)$$

Where E is the elastic modulus, κ is the radius of gyration, A is the cross sectional area, L is length and T is tension. For a circle, $\kappa = d/4$.

Inharmonicity is visible in the higher modes of a string, even when the lower modes agree well with the ideal approximation. The data acquired from the monochord shown in Figure 4.17 shows several dozen clear harmonics. The ratios of these frequencies to those predicted by Equation 4.2 are shown in Figure 4.20. Ideally, all the points would have a value of 1.

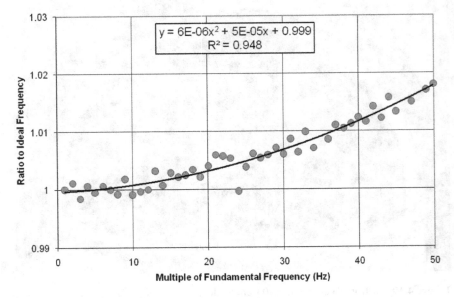

Figure 4.20 String Inharmonicity from Measured Data

4.2.3 Real Strings and Intonation

Strings mounted on a guitar exhibit behavior close to that of an ideal string, but not exactly. Some features of guitars and the sounds they make reflect the fact that the strings do not always behave in an ideal way.

The frequency expression given in Equation (4.2) includes a tension value, T. It is tempting to assume that the tension is constant while the instrument is played, but this is not exactly true. When a string is pressed against a fret as shown in Figure 4.21, it is stretched slightly. Depending on the elastic modulus of the string, the pitch may increase by a noticeable amount; the elastic modulus of steel strings is approximately 30 times that of nylon strings, so this increase in pitch is much more noticeable with steel strings.

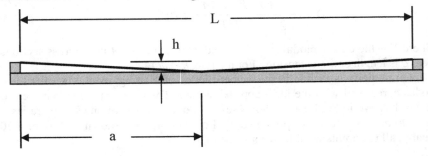

Figure 4.21 String Stretching While Being Played

The length of the stretched string is

$$L + \Delta L = \sqrt{a^2 + h^2} + \sqrt{(L-a)^2 + h^2} \qquad (4.23)$$

The change in tension comes from the definitions of stress and strain

$$\Delta T = EA \frac{\Delta L}{L} = \frac{EA}{L} \left(\sqrt{a^2 + h^2} + \sqrt{(L-a)^2 + h^2} - L \right) \qquad (4.24)$$

And the change in frequency is

$$f + \Delta f = \frac{n}{2L} \sqrt{\frac{T + \Delta T}{\rho}} \qquad (4.25)$$

As an example, consider an E string with $L = 25.5$ in (648 mm), $h = 0.10$ in 2.54 mm), $f = 329.6$ Hz and $a = 12.75$ in (324 mm). This corresponds to a string being pressed against the fretboard at the 12th fret. If the string diameter is 0.010 in (0.25 mm), the nominal tension is 16.3 lb (72.5 N). The change in tension is, $\Delta T = 0.072$ lb (0.322 N). The resulting change in frequency is, $\Delta f = 0.485$ Hz. This change in pitch is small and might be negligible if it was the only deviation from the ideal. However, the bending stiffness of the string can cause a noticeable effect, especially when the string is stopped at higher frets and the effective length is short. The expression for the resonant frequencies of a string with bending stiffness is

$$f_n = \frac{n}{2L} \sqrt{\frac{T}{\rho}} \left[1 + \frac{r^2}{L} \sqrt{\frac{\pi E}{T}} + \left(4 + \frac{n^2 \pi^2}{2} \right) \frac{\pi E r^4}{4 T L^2} \right] \qquad (4.26)$$

Where ρ is the mass per unit length, E is is the elastic modulus of the string material, T is tension, L is length and r is the radius of the string. The expression in braces simply modifies the ideal expression for string resonant frequencies. If the elastic modulus is zero, then the string has no bending stiffness and the result is the ideal expression in Equation (4.2). Anything that makes the expression in the braces larger increases the inharmonicity and anything that makes it smaller decreases inharmonicity. Thus, increasing the radius or elastic modulus makes the deviation from the ideal harmonic series worse, while increasing tension or length makes it better. Note that Equation (4.26) is actually an approximation. The expression in the braces represents the first few terms in a series. The missing terms are assumed to be negligibly small.

Consider now what happens when the string is stopped at the 12th fret. Using the example of the E string above, the change in frequency due to bending stiffness of the string is calculated to be 0.766 Hz. When added to the pitch increase caused by stretching the string, the combined pitch increase is 1.251 Hz. This is small, but large enough to be noticed.

Consider the effect of both stretching and bending on an unwound G string with a resonant frequency of 196 Hz, a diameter of 0.026 in (0.66 mm) and a string tension of 35.3 lb (157 N). This corresponds to a string from a medium weight set of acoustic guitar strings (high E string diameter = 0.013 in, 0.33mm). The pitch increase at the 12th fret due to bending stiffness (inharmonicity) is 2.136 Hz. This is a significant error – larger in both absolute terms and as a percentage – and one that needs to be addressed in the design of the instrument.

The response of guitar designers has been to move the saddle slightly farther from the nut than it would otherwise be. This is only an approximate solution, but it is a relatively easy modification and has proven effective. Most electric guitars are fitted with bridges that have adjustable saddle locations so that they can be changed to accommodate different string weights and height settings. In the vernacular, this is called adjustable intonation. Figure 4.22 shows an electric guitar bridge with adjustable intonation. The individual saddles are not lined up, reflecting the settings required to keep this particular instrument in proper tune.

Setting the intonation on an instrument is a matter of minimizing total error, since a different intonation value is required for each string and for each fret position. A simple approach is to adjust the saddle offset for each string so that it is in tune both when played open and at the 12th fret. While there is still likely to be slight error in pitch at other frets, it should be small. Changing the 'weight' of the strings (their diameters) or the distance of the strings above the fretboard (the action) usually requires adjusting the intonation.

Acoustic guitars are seldom fitted with adjustable bridges. Figure 4.23 shows a non-adjustable bridge on an acoustic guitar. The saddle is not perpendicular to the strings and there is an additional offset on the B string. This is also a common feature.

Figure 4.22 Electric Guitar Bridge with Adjustable Intonation

Figure 4.23 Acoustic Guitar with Fixed Bridge

Note that classical guitars typically have very little intonation in the saddle. A typical instrument may have a saddle set perpendicular to the strings and set 0.090 in (2.29 mm) back from its nominal position. This is possible because the elastic modulus of nylon is about 1/30 of steel. The pitch error due to stretching the strings is very small. Even though the unwound strings have larger diameters than comparable steel strings, the large difference in elastic modulus makes the pitch increase due to bending small.

4.3 Acoustic Guitar Bodies

The sound radiated by an acoustic guitar body is a result of the motion of the surfaces of the body and air motion through the sound hole. These are in turn dominated by forces from the vibrating strings. This is clearly not the whole story, though, or there wouldn't be much variation in the tone of acoustic guitars that were in tune. The dynamic forces from the strings are conditioned by the structural response of the body coupled with the air in and around it. Thus, an understanding of the resonant frequencies and corresponding mode shapes of guitar bodies is important.

While there is, happily, great variation in the design of acoustic guitars, they are generally similar enough that the first two or three modes are similar. The first three modes are shown conceptually in Figure 4.24.

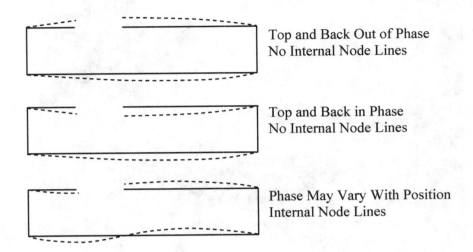

Top and Back Out of Phase
No Internal Node Lines

Top and Back in Phase
No Internal Node Lines

Phase May Vary With Position
Internal Node Lines

Figure 4.24 Conceptual Representation of First Three Guitar Modes

The first mode is sometimes called a breathing mode since the out of phase motion of the top and back causes a net volume change, pushing air in and out of the sound hole as the instrument oscillates. It also correct to call it a 0,0 mode in keeping with the nomenclature established earlier. This mode is often compared to a Helmholtz mode, though a true Helmholtz resonator has rigid walls. The frequency of the first mode is strongly affected by the geometry of the body and the soundhole; the stiffness and mass of the structure have an effect, but it is often secondary. Because the only points of zero displacement are at the edges of the top and back plates, there are no internal node lines.

The second mode is similar to the first in that there are no internal node lines. The primary difference is that the top and back are moving in phase. There is little net volume change and the frequency is more strongly affected by the mechanical properties of the top and back. This is also a 0,0 mode shape, so it is important to clearly distinguish it from the previous mode.

The third mode is the first one in which there are internal node lines. Generalization beyond this is difficult since there is so much variation in the design of acoustic guitars; there may be a net volume change since that depends on the details of the mode shapes and the phase relationship between the top and the back may depend on the location since the node lines on the top and back are not generally in the same places

Figures 4.25a – 4.25d show the mode shapes of a dreadnaught acoustic guitar [72]. They were made using time-averaged holography [73] and the fringe lines connect points with equal displacement. They are analogous to topographic maps used to render geographic features onto a 2D sheet of paper. These images were made using photographic plates and have correspondingly high resolution.

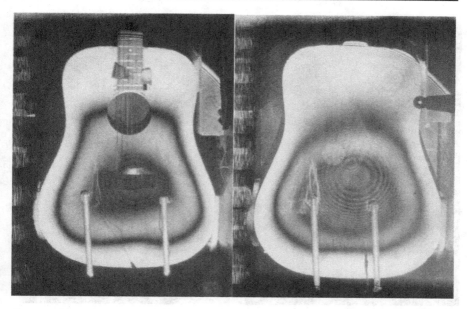

Figure 4.25a Mode Shape of Dreadnaught Guitar at 97Hz.   (Figures 4.14a – 4.14d courtesy, Karl A. Stetson, Karl Stetson Associates, LLC, www.holofringe.com)

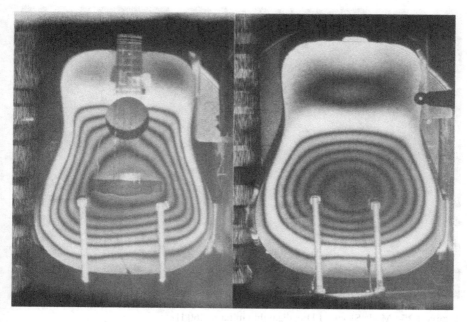

Figure 4.25b Mode Shape of Dreadnaught Guitar at 205Hz

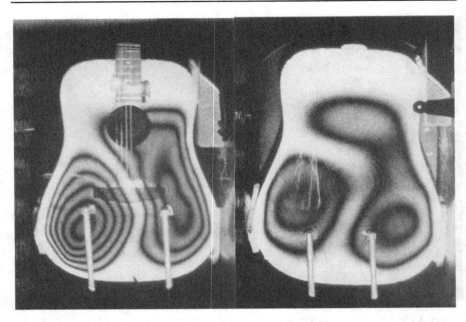

Figure 4.25c Mode Shape of Dreadnaught Guitar at 390 Hz

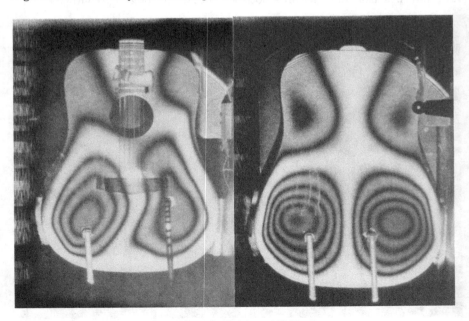

Figure 4.25d Mode Shape of Dreadnaught Guitar at 460 Hz

The geometry of the first two mode shapes is very similar. Note that the difference in the number of fringe lines is means amplitudes are different. If the amplitudes of the left hand side images in Figures 4.25a and 4.25b were normalized so that they had the same number of fringes, they would be nearly identical. The back is structurally quite different from the top, but is coupled through the enclosed air and the sides. Thus, the mode shapes are perhaps more similar than one would expect.

Figures 4.26a – 4.26d show time-averaged holograms of the lower modes of a classical guitar. The geometry and structure of this instrument are different from those in the dreadnaught.

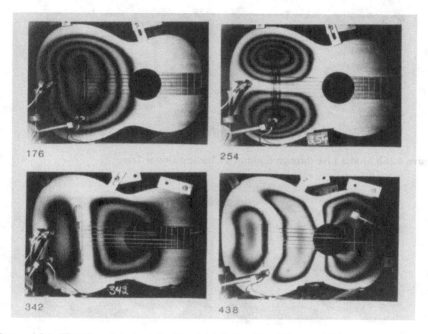

Figure 4.26a First Four Modes of a Classical Guitar Top (Figures 4.26a – 4.26d courtesy, Karl A. Stetson, Karl Stetson Assoc., LLC, www.holofringe.com)

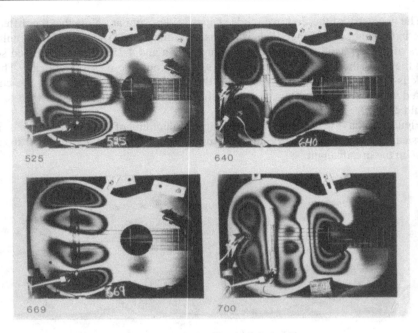

Figure 4.26b Modes Five through Eight of a Classical Guitar Top

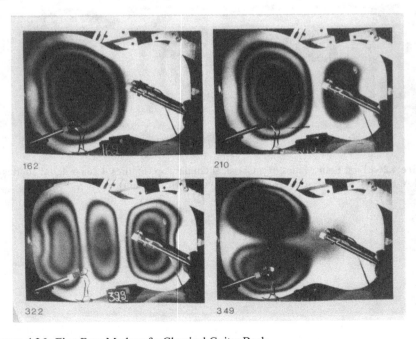

Figure 4.26c First Four Modes of a Classical Guitar Back

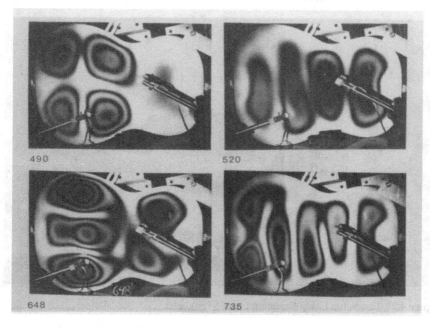

Figure 4.26d Modes Five through Eight of a Classical Guitar Back

Time-averaged holography is an exquisitely sensitive test method and is very useful for musical instruments, where subtle differences in dynamic response can be quite important. However, determining the direction of motion can be difficult – a fringe caused by motion away from the image plane looks just like one formed by motion towards the image plane. Another option for accurately imaging mode shapes is a scanning laser Doppler vibrometer (LDV) as shown in Figure 4.27.

The LDV illuminates a moving surface with a laser and uses the Doppler shift of the reflected beam to determine its velocity. A pair of computer-controlled mirrors allows the beam to scan a surface rather than a single point. To make a full surface measurement, a grid of points is programmed into the control computer and some excitation source is applied. In this case, a small electromagnetic shaker was set to tap the instrument at the right side of the bridge. Input force was measured by a small piezoelectric force sensor specifically designed for tapping. Frequency response functions are measured at each point in the grid and resonant frequencies along with their associate mode shapes are calculated.

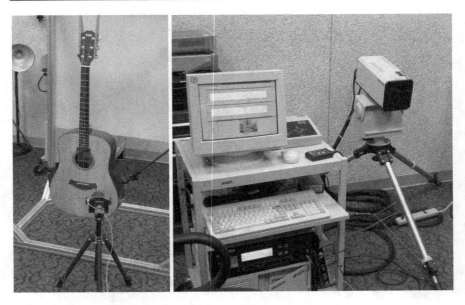

Figure 4.27 Test Setup Using Scanning Laser Vibrometer

Software color codes the velocity results and super-imposes the resulting image over a picture of the target. Figure 4.28 shows results from some of the lower modes of a Taylor Model 710 dreadnaught guitar. The LDV is sensitive to motion along only the beam axis; dark and light areas on the resulting images show velocity in opposite directions.

Figure 4.28a shows the top and back at the first resonant frequency of the body. The point of maximum deflection on both surfaces is roughly in the middle of the lower bout. The top and back are moving out of phase as shown in Figure 4.24, so there is air moving in and out of the sound hole. Note that the diagonal arms appearing in all the figures are there to hold pieces of soft foam against the instrument. This is necessary since the instrument is supported by rubber bands around the tuner knobs and there are several rigid body modes. The foam blocks suppress the rigid body modes without affecting the elastic modes.

Figure 4.28b shows the small electromagnetic shaker used to excite the structure. It was fitted with a small piezoelectric force gauge designed to tap the surface of the structure (rather than being fixed to the structure as is more common). The shaker is driven by the PC that also controls the vibrometer and records the velocity data.

Figures 4.28b shows the third body mode. There are internal node lines present on both the top and back, unlike for the lower body modes. All higher modes are also expected to also have one or more internal node lines. This is particularly important information when finding a location for soundboard pickups. A pickup placed on a node line for a specific mode will not be able to generate a strong signal at that frequency. In acoustic terms, this means that tone will be affected by not having that frequency present.

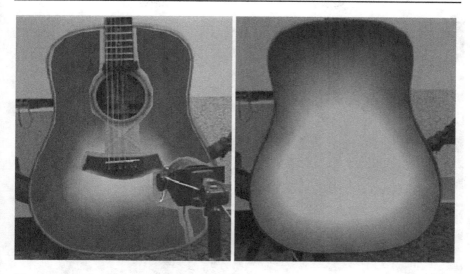

Figure 4.28a Dreadnaught Guitar Top and Back, 99 Hz

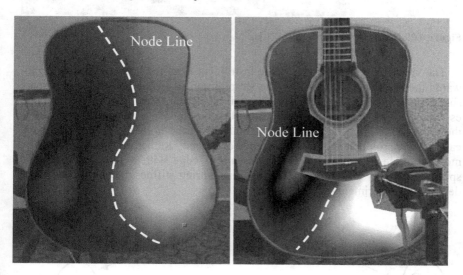

Figure 4.28b Dreadnaught Guitar Top and Back, 321 Hz

The boundary conditions of the top and back are largely a function of how flexible the sides are. The thinner and more flexible the sides are, the closer one would expect to be to a pinned boundary condition. Violins, for example, have sides (called ribs) that are about 1mm thick [57]. Figure 4.29 shows a node line on the side of the guitar shown in Figure 4.28.

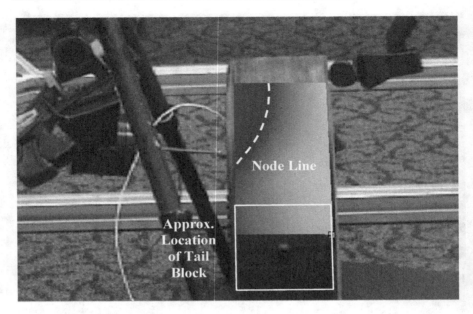

Figure 4.29 Bottom of Guitar Body, 181 Hz

The view in Figure 4.29 shows the instrument from the bottom (looking up towards the neck and headstock). The strap button is visible and the approximate shape of the tail block is outlined. The tail block makes that part of the structure essentially rigid, but there is a clear node line on the side of the instrument above the tail block. This suggests that the boundary conditions for the top and back cannot be approximated as either clamped or pinned. Rather, it would probably be more accurate to model the conditions at the edge with a series of rotational springs as shown in Figure 4.30. The actual spring stiffnesses would need to be determined experimentally.

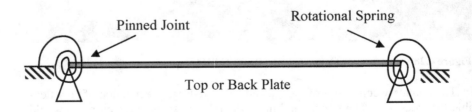

Figure 4.30 Modeling Boundary Conditions for Top and Back

4.4 Complete Instrument

While the dynamic behavior of strings and the body are clearly important, no discussion is complete without considering the response of the complete instrument. The first thing to note is that the entire instrument vibrates when is it played. As an experiment, amplify an acoustic guitar with an accelerometer whose output runs to an electric guitar amplifier rather than the more usual oscilloscope or data acquisition system. Note that piezoelectric accelerometers generally require a signal conditioner and it is the signal conditioner output that runs to the amplifier.

A short time spent moving the accelerometer to different places on the instrument demonstrates clearly that a usable signal can result from points all over the instrument. For example, reasonable sound quality can often be produced from an accelerometer placed at the end of the headstock.

With all the discussion of the resonant frequencies of the instrument, it is easy to forget that the dynamic response of an acoustic instrument is dominated by the string forces. The motion of the instrument and the resulting radiated sound is primarily at the string frequencies and the dynamics of the instrument structure serve to modify the response. Figure 4.31 shows a spectrogram of the audio recording of an open string strum of a steel string acoustic guitar. The recording was made in a high quality hemi-anechoic chamber [74] using a binaural head [75], so the signal is very clean. Note that the acoustic chamber had very low background noise: the noise present was essentially all below 200 Hz. The steady state portion of the signal below 200 Hz is characteristic of this chamber.

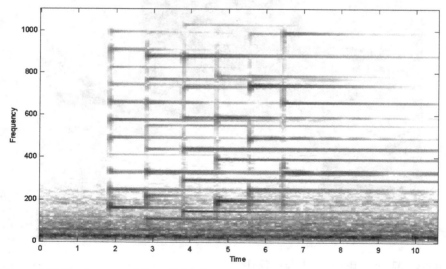

Figure 4.31 Spectrogram of Open String Strum

The recording includes components up to 24 kHz, but, for clarity, only components below 1000 Hz are shown here. The strings were plucked approximately one second apart and it is clear that the fundamental frequency from each string is accompanied by harmonics up to the upper frequency limit of the plot. The time-frequency representation of the data also shows that not all frequencies are present in equal proportion. This is a function of both the point at which the strings were plucked and the dynamic response of the instrument's structure.

Immediately after the first string is plucked, the frequency content of the signal is very simple; it consists of the fundamental frequency of the low E string (ideally 82.4 Hz) and integer multiples of that frequency. By the time the last string is plucked, the signal consists of the fundamental frequency of the high E string (329 Hz) along with the remaining contributions of the other five strings. The frequency content of the signal after plucking the 6th string is structured, but much more richly so than after the first string is played. This frequency structure is responsible for the tone of the instrument

Another characteristic of complete instruments is that they generally have a beam-like bending mode below the first body resonance; frequencies in the range of 50Hz to 80Hz are common. Figure 4.32 shows the results from the vibrometer measurements at 57 Hz. There is a straight node line across the instrument slightly below the waist and, although it isn't visible from this measurement, there is another across the neck above where it crosses the body.

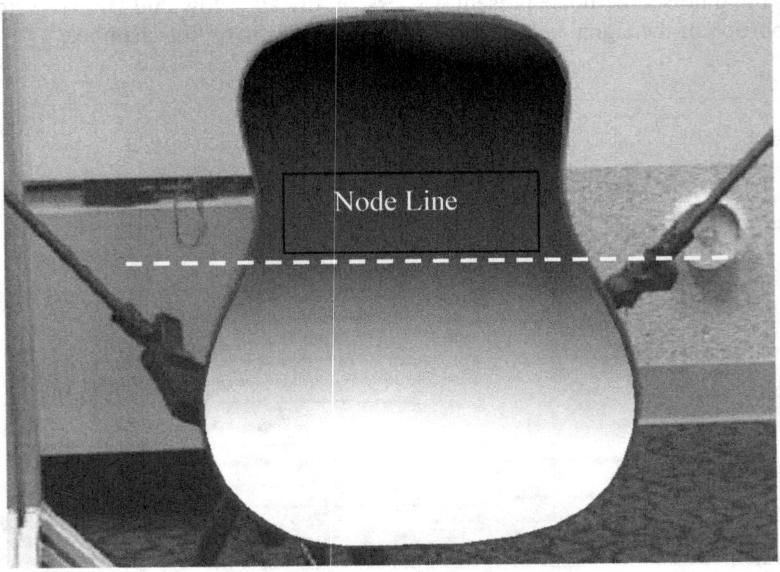

Figure 4.32 Beam Bending Mode at 57 Hz

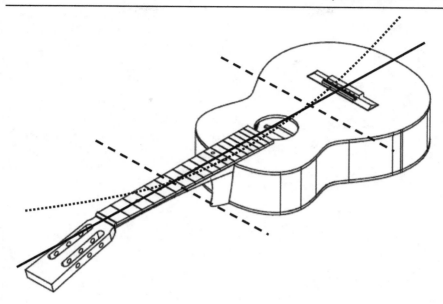

Figure 4.33 Beam Bending Mode on a Complete Guitar

Figure 4.33 shows typical locations of the node lines for the beam bending mode of the instrument along with a reference line showing the un-deformed and deformed shapes of the instrument.

5 Analytical Models

People have been studying the physics of musical instruments at least since the early Renaissance; Galileo, when he wasn't busy with astronomy, optics and upsetting the Catholic Church, studied the motion of a vibrating string [64]. He clearly described the relationship between tension, length and frequency, taking one of the first steps toward putting stringed instrument design on a firm physical foundation.

Historically, stringed instruments have been designed empirically, and progress on describing them using mathematical models was slow until the mid 20th century. Stringed musical instruments pose a surprisingly difficult challenge to the would-be analyst. In order to completely characterize the sound produced, the mathematical model must include structural dynamics, acoustics and the interaction (coupling) between the two. If the results are to include sound quality, the model might also need to account for the physiology of human hearing and the psychoacoustics of how we interpret sound – a pretty tall order indeed.

Since the complete problem is so daunting, people have tended to attack it one piece at a time. Lord Rayleigh's "Theory of Sound" [76], published in 1877 was a landmark work in the field of acoustics and included discussions of music. The technical literature since then has included sporadic works on stringed instruments (mostly violins) up through the middle of the century and these tended to be descriptive – analyzing the sound output or the dynamics of specific instruments [77]. The next step in utility is provided by a predictive model, one that predicts the behavior of an instrument not yet built. Predictive models are often tuned so that their calculated results match the measured results from an instrument. Results of design changes can then be predicted using the tuned model as a reliable starting point.

Attempts to create a mathematical description (a math model) of complete instruments were rare. In retrospect, it is clear that the mathematical tools, in the form of discrete modeling methods and cheap digital computers, were not yet available. An important paper was published in 1962 in which John Schelling described the violin as an equivalent circuit [78]. Digital computers were, at that time, expensive and rare; it made sense to describe a simple analog circuit that could be assembled quickly with a breadboard and readily available components. Schelling's model included both the structure and acoustics and sought to describe the sound produced by the violin by modeling the physics of the instrument. The equivalent circuit models result in mathematical descriptions equivalent to the ones presented here.

R.M. French, *Engineering the Guitar*,
DOI: 10.1007/978-0-387-74369-1_5, © Springer Science+Business Media, LLC 2009

As digital computers and the attendant software became more available, the nature of the models appearing in the technical literature changed. Just to be clear, it makes sense at this point to define a few categories of models. Mathematical models that describe the instrument with a few coupled equations (almost always differential equations) will be called analytical models. These models typically account for only the gross characteristics of the instrument, but have the advantage of being relatively easy to work with. Only the most determined person would try to work with anything more than the smallest analytical model by hand. Rather, they are almost always solved using either a programmable calculator or software on a PC. The analytical models presented here were programmed in Mathcad [79].

At the other end of the mathematical modeling spectrum lie numerical models that describe the instrument by dividing it into smaller elements or a collection of discrete points. While the underlying math is cast in the form of differential equations, numerical models are typically formed by a large number of coupled linear algebraic equations written in matrix form. The most common of this class are finite element models [80]. Finite elements are, approximately speaking, the Lego^{tm} blocks of the math and physics world. The equations that describe the structure of a realistic object like a guitar are far too complex to be solved exactly. However, those describing a simple object like a rectangular brick are much more tractable.

Finite element analysis is built on the idea that simple shapes like rectangles and triangles, which are easy to analyze, can be combined to form more complicated objects. The behavior of the complete structure can be calculated by adding up the contributions of the individual elements from which it is made. The details of this process form an entire field of numerical analysis, but this underlying idea is fairly simple.

5.1 Discrete Models

While the elements contributing to the sound quality of stringed instruments are many and subtle, it is accepted that the first three coupled modes are important [81]. Three elements in the structure of the instrument couple to create these modes: the top of the body, the back of the body and the enclosed volume of air. Figure 5.1 shows the first three modes of a classical guitar as recorded using a scanning laser vibrometer with a small electromechanical shaker providing the excitation signal (the vibrometer uses Doppler shift to measure velocity along the beam axis). The dark and light areas of the instrument are moving in opposite directions and grey areas are not moving. Note that the soundhole is black simply because there is no surface to reflect the laser. The left and center images show the lower bout moving with no node line. The right hand image shows the left and right portions of the lower bout moving out of phase.

Figure 5.1 First Three Modes of a Classical Guitar: 96 Hz, 196 HZ and 294 Hz

Typically, the first two modes are distinguished largely by the phase relationship between the top and the back. In the first mode, the top and back move in opposite directions so that the volume of the body changes. In the second mode they move in the same direction so that there is no nominal change in body volume. The third top mode is roughly anti-symmetric about the centerline of the instrument and is a strong function of the bracing pattern. This is generally the first air-body mode with an internal node line.

Simple discrete analytical structural models have been published which can account for the lowest two or three modes of stringed instruments [82, 83]. An analytical model is just a set of equations that describes some interesting behavior of some physical object like a guitar. These models are very useful for understanding the basic mechanics of the coupled air-structure behavior. Additionally, the effect of structural modifications can be described by calculating the changes in the resonant frequencies predicted by the equations of motion [84]. However, it is important to note that such simple mathematical descriptions cannot describe sound quality of an instrument.

The discrete models presented here are composed of either two or three equations and, thus, use two or three variables. The variables are typically called degrees of freedom (DOF). A degree of freedom can mean different things in different contexts. For our models, a degree of freedom is a mass that moves; the mathematical description that follows will use one variable for each moving part of the instrument. When an analyst talks about working with a 'big' computer model, that usually means one with many degrees of freedom; 100,000 DOF models are not uncommon at this writing.

The equations that form the math models are written in terms of parameters that describe important physical characteristics. In order for the models to be useful, correct values of these parameters must be supplied. This is generally done by selecting a set of parameters so that the predicted resonant frequencies match those measured from an actual instrument. The selection process is often called tuning a model and can take different forms. The simplest models might even be tuned by trial and error – informed guesses at the right values. It is much more likely, though, that the tuning process will depend on some more sophisticated mathematical process intended to minimize the error between the answers predicted by the model and those measured experimentally. Once models are tuned, they can be used to explore the basic physics of the instrument at low frequencies and to predict the effects of design changes.

5.1.1 Two Degree of Freedom Model

The two DOF model couples a flexible top surface with a Helmholtz resonator as shown in Figure 5.2. The walls are assumed to be rigid everywhere except for the flexible section. The flexible section of the top is modeled by a rigid flat plate supported by a linear spring (thus, it cannot have any internal node lines). The two sources of stiffness are this spring and the compressible air inside the body. The two sources of mass are the movable portion of the top and the mass of the air column moving in the sound hole.

The equations of motion for the two DOF model are

$$m_p \ddot{x}_p = F - k_p x_p - R_p \dot{x}_p + A_p \Delta p$$
$$m_h \ddot{x}_h = A_h \Delta p - R_h \dot{x}_h$$

(5.1)

Where m is mass, F is force, k is stiffness, A is area and R is damping. Additionally, Δp is

$$\Delta p = -\mu \left(A_p x_p + A_h x_h \right)$$

(5.2)

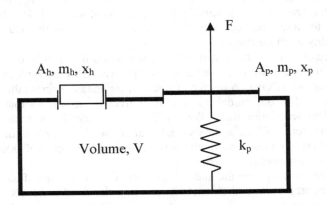

Figure 5.2 Two Degree of Freedom Model

and μ is a proportionality constant between changes in volume and changes in pressure

$$\mu = c^2 \rho / V \qquad (5.3)$$

The model shown in Figure 5.2 is defined in terms of physical parameters. There are:

A_h Area of sound hole
m_h Mass of the air column moving through sound hole
x_h Position of air column moving through sound hole
V Volume of enclosed air
k_p Stiffness of top plate
F Force applied to top plate
A_p Effective area of top plate
m_p Effective mass of top plate
x_p Position of top plate

In terms of these parameters, the equations of motion become

$$m_p \ddot{x}_p = F - \left(k_p + \mu A_p^2 \right) x_p - R_p \dot{x}_p - \mu A_h A_p x_h$$
$$m_h \ddot{x}_h = -\mu A_h^2 x_h - R_h \dot{x}_h - \mu A_h A_p x_p \qquad (5.4)$$

In matrix form, the equations of motion are

$$\begin{bmatrix} m_p & 0 \\ 0 & m_h \end{bmatrix} \begin{Bmatrix} \ddot{x}_p \\ \ddot{x}_h \end{Bmatrix} + \begin{bmatrix} R_p & 0 \\ 0 & R_h \end{bmatrix} \begin{Bmatrix} \dot{x}_p \\ \dot{x}_h \end{Bmatrix} + \begin{bmatrix} k_p + \mu A_p^2 & \mu A_h A_p \\ \mu A_h A_p & \mu A_h^2 \end{bmatrix} \begin{Bmatrix} x_p \\ x_h \end{Bmatrix} = \begin{Bmatrix} F \\ 0 \end{Bmatrix} \qquad (5.5)$$

This equation is a second order ordinary differential equation similar to those that appear in all texts on mechanical vibrations. The mass, damping and stiffness matrices are usually called C and K respectively, so the equations are written as

$$[M]\{\ddot{x}\} + [C]\{\dot{x}\} + [K] = \{F\} \qquad (5.6)$$

With the equations of motion at hand, there are basically two related ways of extracting useful results. The first is to calculate eigenvalues (resonant frequencies of the guitar) and the second is to calculate a frequency response functions for comparison with test data. In either case, the equation must be cast in frequency domain.

$$\left[-\omega^2\begin{bmatrix} m_{\mathrm{p}} & 0 \\ 0 & m_{\mathrm{h}} \end{bmatrix}+i\omega\begin{bmatrix} R_{\mathrm{p}} & 0 \\ 0 & R_{\mathrm{h}} \end{bmatrix}+\begin{bmatrix} k_{\mathrm{p}}+\mu A_{\mathrm{p}}^2 & \mu A_{\mathrm{h}} A_{\mathrm{p}} \\ \mu A_{\mathrm{h}} A_{\mathrm{p}} & \mu A_{\mathrm{h}}^2 \end{bmatrix}\right]\left\{\begin{matrix} X_{\mathrm{p}} \\ X_{\mathrm{h}} \end{matrix}\right\}=\left\{\begin{matrix} F_{\mathrm{p}}(\omega) \\ 0 \end{matrix}\right\} \quad (5.7)$$

If we assume the damping terms R_{p} and R_{h} are reasonably small, they can be omitted when we calculate the resonant frequencies of the instrument. The eigenvalue calculation also assumes there is no external force (free response). The eigenvalue equation is then

$$\left[\begin{bmatrix} k_{\mathrm{p}}+\mu A_{\mathrm{p}}^2 & \mu A_{\mathrm{h}} A_{\mathrm{p}} \\ \mu A_{\mathrm{h}} A_{\mathrm{p}} & \mu A_{\mathrm{h}}^2 \end{bmatrix}-\omega^2\begin{bmatrix} m_{\mathrm{p}} & 0 \\ 0 & m_{\mathrm{h}} \end{bmatrix}\right]\left\{\begin{matrix} X_{\mathrm{p}} \\ X_{\mathrm{h}} \end{matrix}\right\}=\left\{\begin{matrix} 0 \\ 0 \end{matrix}\right\} \quad (5.8)$$

where ω is the natural frequency in rad/sec. The equation is always satisfied if X_{p} and X_{h} are both zero, but this is not a very interesting result. Not a group to mince words, mathematicians call this the trivial solution. The only useful answers result when the vector of displacements is non-zero. In that case, the determinant of the matrix must be zero.

$$\left|\begin{bmatrix} k_{\mathrm{p}}+\mu A_{\mathrm{p}}^2 & \mu A_{\mathrm{h}} A_{\mathrm{p}} \\ \mu A_{\mathrm{h}} A_{\mathrm{p}} & \mu A_{\mathrm{h}}^2 \end{bmatrix}-\omega^2\begin{bmatrix} m_{\mathrm{p}} & 0 \\ 0 & m_{\mathrm{h}} \end{bmatrix}\right|=\left|\begin{bmatrix} k_{\mathrm{p}}+\mu A_{\mathrm{p}}^2-\omega^2 m_{\mathrm{p}} & \mu A_{\mathrm{h}} A_{\mathrm{p}} \\ \mu A_{\mathrm{h}} A_{\mathrm{p}} & \mu A_{\mathrm{h}}^2-\omega^2 m_{\mathrm{h}} \end{bmatrix}\right|=0 \quad (5.9)$$

There are two values of ω that will satisfy Equation (5.8). These are the two resonant frequencies of the instrument. The corresponding values of X_{p} and X_{h} form the eigenvectors (the modes). Fortunately, there is a closed form solution for the 2-DOF eigenvalue problem.

$$\left(k_{\mathrm{p}}+\mu A_{\mathrm{p}}^2-\omega^2 m_{\mathrm{p}}\right)\left(\mu A_{\mathrm{h}}^2-\omega^2 m_{\mathrm{h}}\right)-\left(\mu A_{\mathrm{h}} A_{\mathrm{p}}\right)\left(\mu A_{\mathrm{h}} A_{\mathrm{p}}\right)=0 \quad (5.10)$$

The eigenvalue usually is denoted by λ, so $\lambda=\omega^2$. Expanding out the polynomial gives

$$m_{\mathrm{p}}m_{\mathrm{h}}\lambda^2-\left(k_{\mathrm{p}}m_{\mathrm{h}}+\mu A_{\mathrm{p}}^2 m_{\mathrm{h}}+m_{\mathrm{p}}\mu A_{\mathrm{h}}^2\right)\lambda+\left(k_{\mathrm{p}}\mu A_{\mathrm{h}}^2+\mu^2 A_{\mathrm{p}}^2 A_{\mathrm{h}}^2-\mu^2 A_{\mathrm{h}}^4\right)=0 \quad (5.11)$$

The roots of this quadratic equation are the eigenvalues of the 2-DOF model. If the flexible element is removed, then $x_{\mathrm{p}}=0$ and the result is a Helmholtz resonator. The resonant frequency is $\omega_{\mathrm{h}}=(\mu A_{\mathrm{h}}^2/m_{\mathrm{h}})^{1/2}$.

5.1.2 Two DOF Example

A 2-DOF model was tuned to match the response of a Taylor steel string grand auditorium style guitar as shown in Figure 5.3.

The model tuning process is described in detail in a later section. For now, the model parameters are given. Tests showed that the first two resonant frequencies of the instrument were 99.4 Hz and 185 Hz and the Helmholz frequency was 115 Hz. By comparing test data with calculated frequencies, the model parameters were determined to be those shown in Table 5.1. The calculated resonance frequencies using this model are 99.31 Hz and 183.8 Hz. The calculated Helmholtz frequency is 114.3 Hz.

The model predicts resonant frequencies assuming there is no damping. Clearly, there is energy loss – the fact that the instrument radiates sound means that it is radiating away energy. Damping can be estimated by comparing calculated and measured frequency response functions.

Figure 5.3 Taylor Grand Auditorium Acoustic Guitar (image courtesy Taylor Guitars, www.taylorguitars.com)

Table 5.1 Parameters for 2-DOF Model

Parameter	Description	Value
A_h	Sound hole area	81.89 cm^2
A_p	Top plate area	630.0 cm^2
c	Speed of sound in air	338 m/s
k_p	Top stiffness	191,400 N/m
m_h	Mass of air column in sound hole	1.247 gm
m_p	Top mass	190.1 gm
V	Volume of body	14.66 L
ρ	Density of air	1.23 kg/m^3

The frequency response function (FRF) is the ratio of output to input in the frequency domain. Frequency response functions are also called transfer functions and the two terms are often used interchangeably. There are two degrees of freedom in the analytical model and, thus, two possible outputs and two possible inputs. There are then four possible FRFs. For reasons that remain a mystery, frequency response functions are almost always denoted by H. We are interested in the FRF at the top plate due to the force on the top plate, $H_{pp}=X_p/F_p$.

The equations of motion in frequency domain are

$$\left[-\omega^2\begin{bmatrix} m_p & 0 \\ 0 & m_h \end{bmatrix} + i\omega\begin{bmatrix} R_p & 0 \\ 0 & R_h \end{bmatrix} + \begin{bmatrix} k_p + \mu A_p^2 & \mu A_h A_p \\ \mu A_h A_p & \mu A_h^2 \end{bmatrix}\right]\left\{\begin{matrix} X_p \\ X_h \end{matrix}\right\} = \left\{\begin{matrix} F_p(\omega) \\ F_h(\omega) \end{matrix}\right\} \quad (5.12)$$

An FRF is the ratio of output to input, so the equations can be re-written

$$\left\{\begin{matrix} X_p \\ X_h \end{matrix}\right\} = \left[-\omega^2\begin{bmatrix} m_p & 0 \\ 0 & m_h \end{bmatrix} + i\omega\begin{bmatrix} R_p & 0 \\ 0 & R_h \end{bmatrix} + \begin{bmatrix} k_p + \mu A_p^2 & \mu A_h A_p \\ \mu A_h A_p & \mu A_h^2 \end{bmatrix}\right]^{-1}\left\{\begin{matrix} F_p(\omega) \\ F_h(\omega) \end{matrix}\right\} \quad (5.13)$$

In shorthand form, this is

$$\left\{\begin{matrix} X_p \\ X_h \end{matrix}\right\} = \begin{bmatrix} H_{pp} & H_{ph} \\ H_{hp} & H_{hh} \end{bmatrix}\left\{\begin{matrix} F_p(\omega) \\ F_h(\omega) \end{matrix}\right\} \quad (5.14)$$

Writing out the two equations separately gives

$$X_p = H_{pp}F_p + H_{ph}F_h$$
$$X_h = H_{hp}F_p + H_{hh}F_h \quad (5.15)$$

It's clear that [H] is the transfer function matrix. The only force on the instrument is at the top plate – the strings acting through the saddle and bridge – so the relationship between the input force and the output is

$$X_p = H_{pp}F_p$$
$$X_h = H_{hp}F_p \quad (5.16)$$

To calculate an FRF, make a list of frequencies and calculate a transfer function matrix for each frequency. The (1,1) term of each of these matrices form the FRF, H_{pp}. Figure 5.4 shows the calculated FRF when $R_h=R_p=0.01$ N·s/m. Note that this FRF is calculated as position of the top plate per unit force. Measurements are more often performed in terms of acceleration per unit force. The two peaks are the coupled resonant frequencies and the anti-resonance between them is the Helmholtz frequency of the enclosed air volume.

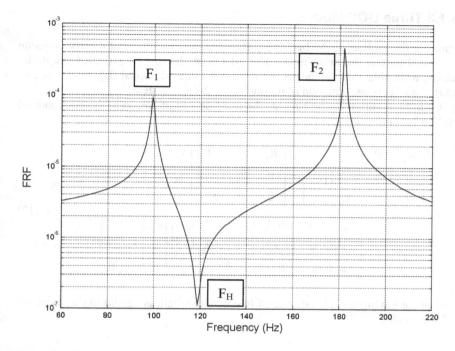

Figure 5.4 Calculated Frequency Response Function for Top Plate Showing First Two Coupled Air-Body Resonant Frequencies and Helmholtz Frequency

It's important at this point to catch up on some technical lingo. The transfer function can be based on position, velocity or acceleration and each has its own name as shown in Table 5.2. While the names are arcane, they are precise and appear regularly in the technical literature on musical instruments.

Table 5.2 Transfer Function Nomenclature

Response Type	Transfer Function Name	Mathematical Form
Position	Receptance	$\dfrac{X(\omega)}{F(\omega)} = H(\omega) = \left[-\omega^2 M + i\omega C + K\right]^{-1}$
Velocity	Mobility	$\dfrac{i\omega X(\omega)}{F(\omega)} = i\omega H(\omega) = i\omega\left[-\omega^2 M + i\omega C + K\right]^{-1}$
Acceleration	Accelerance	$\dfrac{-\omega^2 X(\omega)}{F(\omega)} = -\omega^2 H(\omega) = -\omega^2\left[-\omega^2 M + i\omega C + K\right]^{-1}$

5.1.3 Three DOF Model

Ignoring the motion of the back simplifies the mathematics, but clearly omits an important part of the instrument. The 2-DOF model can be extended simply by adding a flexible back as shown in Figure 5.5. The additional degree of freedom adds a third coupled equation of motion to the 2-DOF system

With the addition of the back as the third degree of freedom, the equations of motion become

$$
\begin{bmatrix} m_p & 0 & 0 \\ 0 & m_h & 0 \\ 0 & 0 & m_b \end{bmatrix} \begin{Bmatrix} \ddot{x}_p \\ \ddot{x}_h \\ \ddot{x}_b \end{Bmatrix} + \begin{bmatrix} R_p & 0 & 0 \\ 0 & R_h & 0 \\ 0 & 0 & R_b \end{bmatrix} \begin{Bmatrix} \dot{x}_p \\ \dot{x}_h \\ \dot{x}_b \end{Bmatrix} +
$$

$$
\begin{bmatrix} k_p + \mu A_p^2 & \mu A_h A_p & \mu A_p A_b \\ \mu A_h A_p & \mu A_h^2 & \mu A_b A_h \\ \mu A_p A_b & \mu A_b A_h & k_b + \mu A_b^2 \end{bmatrix} \begin{Bmatrix} x_p \\ x_h \\ x_b \end{Bmatrix} = \begin{Bmatrix} F \\ 0 \\ 0 \end{Bmatrix}
$$

(5.17)

Except for having the third degree of freedom, the math is similar to that from the 2-DOF model.

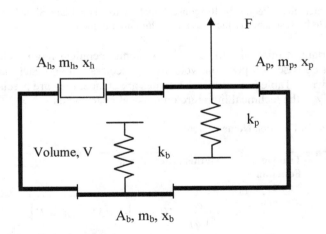

Figure 5.5 Three Degree of Freedom Model

5.1.4 Three DOF Example

Before the discrete model can be of much use, it must be tuned to match the instrument in question. Several of the parameters in the discrete model represent effective values that can't be directly measured from the instrument. Rather, they are inferred from measurements of the instrument's natural frequencies. In mathematical terms, this is an inverse problem. There is a set of equations with unknown parameters that predict the dynamic response of an instrument. Using data from a properly designed test, it is possible to identify the parameters so that the analytical model gives results that match experimental results.

The typical test uses a single sensor, usually an accelerometer, fixed to the soundboard of the instrument and an instrumented hammer tapping on the soundboard as close to the accelerometer as is practical (Figure 5.6). The hammer is fitted with a force gauge so that the input force can be measured simultaneously with the acceleration response. The instrument is placed on soft supports such as foam blocks (to isolate it for the structure of the table) and the strings are damped, again using foam. The FRF is calculated by averaging the results of several impacts. In practice, five impacts are usually enough. The resulting FRF (Figure 5.7) shows the first three natural frequencies that should be matched by the math model. The problem is that there are more unknown parameters in the 3-DOF math model than there are natural frequencies. Thus, there can be many combinations of numbers that will satisfy the equations.

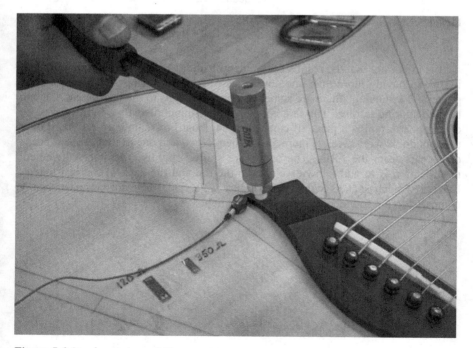

Figure 5.6 Accelerometer and Hammer

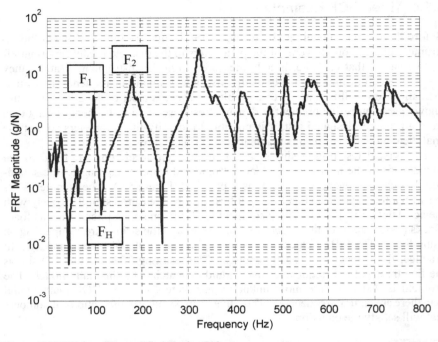

Figure 5.7 FRF from Unmodified Taylor 710

It is important that the results of the identification process are physically mean-ingful, so having more than one unique set of measured data would improve the process. Fortunately, there are some easy modifications to the instrument that can be made during testing. The first is to cover the soundhole with a light, stiff piece of cardboard; taping a piece of corrugated cardboard or foam core art board over the soundhole works well (Figure 5.8). Then $A_h = 0$ and one of the resonant peaks disappears from the FRF. The others are shifted slightly. Another convenient modification is to add known weights to the top and the back of the instrument. Ratchet wrench sockets work well (Figure 5.9) since they are compact, heavy compared to the soundboard and readily available.

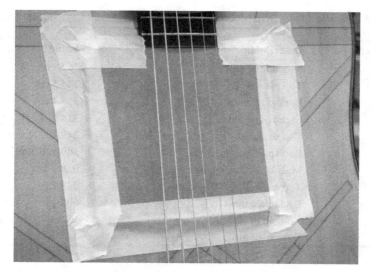

Figure 5.8 Soundhole Cover Made From Corrugated Cardboard

Figure 5.9 Socket Used As a Soundboard Mass

The ideal data set includes response measurements from all the possible test configurations. Once the test is set up, the additional configurations require very little additional time. There are 21 measured data points to be used for identifying the nine unknown parameters (excluding damping values). Since this is an over defined problem, it is much less likely that a physically unrealistic set of parameters will be identified. Table 5.3 shows the model parameters found to minimize error between the measured and calculated frequencies. Table 5.4 shows data acquired from a Taylor Model 710 dreadnought guitar.

Table 5.3 Parameters for 3-DOF Model

Parameter	Description	Value
A_h	Sound hole area	76.8 cm^2
A_p	Top plate area	748.5 cm^2
A_b	Back plate area	556.4 cm^2
c	Speed of sound in air	338 m/s
k_p	Top stiffness	248,400 N/m
k_b	Back plate stiffness	180,500 N/m
m_h	Mass of air column in sound hole	0.967 gm
m_p	Top mass	248.7 gm
m_b	Back plate mass	60.0 gm
V	Volume of body	15.5 L
ρ	Density of air	1.23 kg/m^3

Table 5.4 Measured Frequencies for Tuning 3-DOF Discrete Model (Hz)

Description	Exp.			Calc.		
	f_1	f_2	f_3	f_1	f_2	f_3
Nominal	98.1	181.9	325.6	96.5	177.6	302.5
Soundhole Open, Top Mass (46.2 gm)	96.3	170.6	293.1	95.5	165.2	302.1
Soundhole Open, Back Mass (51.3 gm)	97.2	182.5	325.0	95.9	173.1	229.6
Soundhole Open, Top and Back Mass	95.2	171.2	292.2	94.9	161.8	227.8
Soundhole Covered	–	167.5	320.9	–	171.5	298.4
Soundhole Covered, Top Mass	–	155.0	284.4	–	157.7	298.0
Soundhole Covered, Back Mass	–	160.6	182.1	–	169.4	221.8
Soundhole Covered, Top and Back Mass	–	150.9	180.0	–	156.5	220.6
	Helmholz Frequency,	113.5 Hz		118.2		

Static testing can also be used to estimate the stiffness of the top and back of the instrument. A simple fixture can be used to measure the top and back deformation with a known weight and a dial gauge. The ratio of the weight to the deformation is the stiffness. The discrete model uses an effective stiffness that may not exactly match an experimental value. However, the measured data can be used as a starting point for identifying the unknown parameters in the equations of motion.

5.1.5 Calculating the Effects of Design Changes

One useful application of the simple discrete models is predicting the effect of design changes on the natural frequencies [84]. The analytical method of predicting the effect of design changes is to use partial derivatives; it is possible to develop an expression that predicts the change in the natural frequencies as one of the model parameters is changed. A simpler approach is to just vary parameters in a tuned model and to note the effects on the calculated resonant frequencies. Let's consider the effects of some design changes on the predicted behavior of an acoustic guitar. Table 5.5 shows some predicted effect of some simple design changes.

Table 5.5 Predicted Effect of Design Changes Using 3-DOF Model

Design Change	Δf_1 (Hz)	Δf_1 (%)	Δf_2 (Hz)	Δf_2 (%)
Nominal	0	0	0	0
Increase top mass 20%	−1.13	−1.17	−13.3	−7.48
Increase top Stiffness 20%	2.37	2.46	12.2	6.84
Increase Soundhole Area 20%	15.9	16.4	4.14	2.33
Increase Body Volume 20%	−5.28	−5.46	−3.17	−1.69

Note that the results presented here are for 20% increases in the model parameters. Decreases in the model parameters can be assumed to have the opposite effect.

If you wish to calculate the effect of design changes (sometimes called design sensitivities) in analytical terms, there are two basic approaches. The first is to develop the expressions for eigenvalue derivatives directly from the matrix form of the equations of motion [83, 84]. The second is to write out an expression for the natural frequencies and then find the derivatives with respect to changes in the model parameters. The first approach is more general, but beyond the scope of this work, but the second is relatively compact and easy to describe here.

For the 2-DOF model, it is possible to write out a closed form expression for the eigenvalues. The matrix form of the equations of motion is

$$\left[\begin{bmatrix} k_{11} & k_{12} \\ k_{21} & k_{22} \end{bmatrix} - \lambda \begin{bmatrix} m_{11} & 0 \\ 0 & m_{22} \end{bmatrix} \right] \begin{Bmatrix} x_1 \\ x_2 \end{Bmatrix} = 0 \tag{5.18}$$

This is just the eigenvalue equation for the 2-DOF model as shown in Equation (5.8). Note that $\lambda = \omega^2$; this is standard notation in the technical literature. There are two values of λ that correspond to the two resonant frequencies. Using some basic ideas from matrix algebra, the two values of λ are the roots of an expression called the characteristic equation. It is just a quadratic polynomial of the form

$$\left(k_{11} - \lambda m_{11} \right)\left(k_{22} - \lambda m_{22} \right) - k_{12}k_{21} = 0$$
$$m_{11}m_{22}\lambda^2 - \left(k_{11}m_{22} + k_{22}m_{11} \right)\lambda + \left(k_{11}k_{22} - k_{12}k_{21} \right) = 0 \tag{5.19}$$

The eigenvalues, λ_1 and λ_2, can be written out explicitly using the quadratic equation

$$\lambda_{1,2} = \frac{k_{11}m_{22} + k_{22}m_{11} \pm \sqrt{\left(k_{11}m_{22} + k_{22}m_{11}\right)^2 - 4m_{11}m_{22}\left(k_{11}k_{22} - k_{12}k_{21}\right)}}{2m_{11}m_{22}} \quad (5.20)$$

The eigenvalue sensitivity is just the derivative of this expression with respect to some model parameter. For example, the effective mass of the soundboard is m_{11}. Since $m_{11} = m_p$, the sensitivity is just the derivative if Equation (5.20) with respect to m_p. Finally, the change in the eigenvalues due to a change in top mass is

$$\Delta\lambda_1 = \frac{\partial\lambda_1}{\partial m_{11}}\Delta m_{11}$$

$$(5.21)$$

$$\Delta\lambda_2 = \frac{\partial\lambda_2}{\partial m_{11}}\Delta m_{11}$$

5.2 Finite Difference Models

Finite difference (FD) models occupy a middle space between simple discrete models and finite element models [85]. It is well beyond the capability of all but the most trained analysts to develop a finite element model of a guitar using anything other than a readily available software package. Finite difference models, however, can be created directly from the governing equations using a simple approximation process and don't necessarily require advanced software. As such, they offer the opportunity to produce more accurate simulations of the structural acoustics of guitars without imposing unmanageable modeling and software requirements. The discussion here closely follows the development presented by Kristiansen, Dhainuaut and Johansen in Chapter 5 of reference [85]; the reader is encouraged to refer back to it for numerical results and methods for improving the accuracy.

FD models can be 2-D or 3-D. Here, the discussion is limited to 2-D models. The additional complication of going to a three dimensional model is such that the effort might be better spent on a finite element model.

5.2.1 Finite Difference Acoustic Models

The starting point for finite difference acoustic models is the wave equation. In its most general form, the wave equation is written in terms of a potential, ϕ.

$$\nabla^2\phi + k\ddot{\phi} = 0 \quad (5.22)$$

For the 2-D acoustics problems, the potential is pressure. Transforming the resulting wave equation into frequency domain gives the Helmholtz equation

$$\frac{\partial^2 p}{\partial x^2} + \frac{\partial^2 p}{\partial y^2} + \frac{\omega^2}{c^2} p = 0 \tag{5.23}$$

Where c is the speed of sound

The core idea underlying the development of FD models is that partial derivatives can be replaced with finite difference approximations based on a model with a finite number of grid points. The space in which the Helmholtz equation is to be solved is divided into a rectangular grid where the location of each grid point is (x_i, y_j).

There are two forms of finite difference approximations to a derivative, the forward difference and the backward difference. If the grid spacing is regular, then $x_{i+1} - x_i = x_i - x_{i-1} = \Delta x$.

$$\frac{\partial p_{i,j}}{\partial x_{i,j}} \approx \frac{p_{i+1,j} - p_{i,j}}{x_{i+1,j} - x_{i,j}} = \frac{p_{i+1,j} - p_{i,j}}{\Delta x} \quad \text{(forward)}$$

$$\frac{\partial p_{i,j}}{\partial x_{i,j}} \approx \frac{p_i - p_{i-1}}{x_{i,j} - x_{i-1,j}} = \frac{p_i - p_{i-1}}{\Delta x} \quad \text{(backward)} \tag{5.24}$$

The second derivative requires three points rather than two. It is convenient to use one point on each side of the (i,j) point; this is done by combining the forward and the backward differences to form a central difference.

$$\frac{\partial^2 p_{i,j}}{\partial x_{i,j}^2} \approx \frac{1}{x_{i,j} - x_{i-1,j}} \left[\frac{\partial p_{i,j}}{\partial x_{i,j}} - \frac{\partial p_{i-1,j}}{\partial x_{i-1,j}} \right]$$

$$\approx \frac{1}{\Delta x} \left[\frac{p_{i+1,j} - p_{i,j}}{\Delta x} - \frac{p_{i,j} - p_{i-1,j}}{\Delta x} \right] = \frac{1}{\Delta x^2} \left[p_{i+1,j} - 2p_{i,j} + p_{i-1,j} \right] \tag{5.25}$$

Derivatives with respect to y are formed the same way. Substituting this expression in the Helmholtz equation gives, for point (i,j)

$$\frac{\partial^2 p_{i,j}}{\partial x_{i,j}^2} + \frac{\partial^2 p_{i,j}}{\partial y_{i,j}^2} + \frac{\omega^2}{c^2} p_{i,j} = 0$$

$$\approx \frac{1}{\Delta x^2} \left[p_{i+1,j} - 2p_{i,j} + p_{i-1,j} \right] + \frac{1}{\Delta y^2} \left[p_{i,j+1} - 2p_{i,j} + p_{i,j-1} \right] + \frac{\omega^2}{c^2} p_{i,j} = 0 \tag{5.26}$$

The finite difference approximation converts the Helmholtz equation from a single partial differential equation to a group of coupled algebraic equations – there will be one equation for every grid point. Before the set of equations can be solved, boundary conditions must be applied. For hard walls, the pressure boundary condition is

$$\frac{\partial p}{\partial n} = 0 \tag{5.27}$$

That is, the slope of the pressure normal to the hard wall is zero. Thus, for vertical walls, $\partial p/\partial x = 0$ and for horizontal walls, $\partial p/\partial y = 0$. Since the derivatives are approximated by finite differences, the hard wall boundary condition requires a point on the wall and one more adjacent to it. Rather than defining the boundary condition using internal points, artificial external points are created as shown Figure 5.10.

The next step in developing the model is to add one or more flexible walls. One approach is to treat one or more walls as a simply supported beam. The differential equation describing behavior of a beam is

$$EI\frac{\partial^4 \eta}{\partial x^4} + \rho\omega^2\eta + p = 0 \tag{5.28}$$

Where ρ is density, η is displacement and E is the elastic modulus of the beam. I is the area moment of inertia, a number that specifies beam stiffness due to its cross-sectional shape. For a rectangular cross section, $I = bh^3/12$ where b is the width and h is the height. In practice, EI can be treated as a single number that specifies stiffness of the beam. Simple supports at the ends of the beam require that the displacement and moment be zero at the ends of the beam. Zero moment implies zero curvature, so the end conditions are

$$\eta(0) = \eta(L) = \left.\frac{\partial^2 \eta}{\partial x^2}\right|_{x=0} = \left.\frac{\partial^2 \eta}{\partial x^2}\right|_{x=L} = 0 \tag{5.29}$$

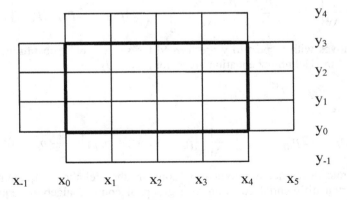

Figure 5.10 Hard-Walled Cavity

The finite difference approach used to discretize Equation (5.19) is applied to the beam equation and the result is the addition of unknown values of η for every unconstrained grid point on the flexible sections of the wall. Thus, the number of variables is equal to the number of unknown pressures plus the number of unknown displacements along the flexible surfaces. The interface between the flexible wall and the enclosed air is addressed by a boundary condition requiring that the particle velocity of the air at the moving wall match that of the wall.

$$i\omega\eta = -\frac{1}{i\omega\rho_a}\frac{\partial p}{\partial y}$$
(5.30)

Where ρ_a is the air density (nominally 1.23 kg/m^3) at sea level. The partial derivative with respect to y assumes the flexible surfaces are at the top and bottom of the enclosed cavity. If they were at the ends, the partial derivative would be with respect to x.

With both acoustic and structural models in hand, the only remaining step is to combine them into a coupled structural acoustics model. This is mainly a matter of enforcing the proper boundary conditions at the interface. At the sound hole, the pressure is zero ($p_{i,j} = 0$).

5.3 Finite Element Models

Calculating the displacements of any structure requires solving either integral or differential equations. The difficulty of solving them is a function of geometry – simple shapes are much easier to analyze than complex ones. The problem is that even relatively simple structures can still be too complicated to analyze exactly. Something as simple as a parallelogram-shaped plate with a square hole placed slightly off-center poses a nearly impossible analytical challenge. This is where finite elements come in.

Finite elements are sort of the Lego^{tm} blocks of the mathematical world. It is relatively easy to model the behavior of a simple block of material. If there was a way to mathematically assemble collections of these simple blocks to approximate complex shapes, it would be possible to analyze just about anything. This is exactly what the finite element method does. Finite element modeling software typically contains a preprocessor to make the model using library of simple elements, a computational engine to do the actual number crunching and a postprocessor to present the results in a useful graphical form.

Figure 5.11 shows a finite element model of a solid body electric guitar that has been color coded to show the von Mises stress [86]. It is not surprising to note that the stresses in the body are generally low and that the location of maximum stress is the nut. The nut is where the strings are bent through an angle of about 15° and the neck is the thinnest. The actual elements are visible in much of the model and it's clear that the instrument was modeled using tetrahedral elements.

Figure 5.11 A Finite Element Model of a Solid Body Guitar (image courtesy MSC Software, www.mscsoftware.com)

Acoustic guitars present a special analytical challenge because the vibrating structure of the instrument is closely coupled with the air both in and around the instrument. Many finite element packages include acoustic elements that can be used to model the air. Modeling this volume necessarily means adding a large number of additional variables (degrees of freedom). The benefit of dealing with the increased number of variables is that pressures can be calculated at points away from the instrument.

5.4 Boundary Element Models

An alternative to finite element modeling is boundary element modeling. Boundary elements are not yet as developed as are finite elements, but offer great promise for modeling acoustic structural interactions. The boundary element method casts the governing differential equations in integral form for solution [87]. It also employs basic principles of vector calculus (Green's function) to transform a volume integral into a surface integral [88].

A primary advantage of the boundary element method is that is requires discretizing only the boundary of the volume of interest. In contrast, a finite element model of an instrument and the associated air would require discretizing the entire volume of interest and the number of elements would be correspondingly larger. This is particularly a problem if the desire is to calculate far field radiation properties. The boundary element can deal with exterior problems as easily as interior problems.

There are issues of numerical efficiency with boundary element method. Finite element models typically generate large matrices, but they are, symmetric, almost always sparsely populated and can be structured to have low bandwidth. Thus, efficient, specialized methods can be used to invert them or to extract eigenvalues. Boundary element formulations result in fully populated matrices without these simplifying characteristics. Even though the number of equations may be significantly smaller, the computational load is not proportionately lower.

5.5 Geometry Models

Math models discussed so far are all intended to predict dynamic behavior of the guitar structure coupled with the air in and around it. However, it is important to understand that modeling the geometry of the instrument is also important. Geometric modeling is often lumped under the term Computer Aided Design (CAD). In order to use computer controlled machines to manufacture parts or to generate accurate finite element models, it is first necessary to have an accurate geometric description of the instrument.

Instruments can only be as precise as the instructions that are driving the machines. Even the best books on guitar making offer only the most basic description of the shape. It's not unusual for the instructions to start with something like, "Draw a straight line on a sheet of brown wrapping paper to use as a centerline."

Traditionally, luthiers have cut body templates from plywood or plastic. However, as computer-controlled milling equipment gets more accessible, the hand-cut template is increasingly insufficient. Clearly, it would be helpful to develop a closed form expression to describe the body shape so that precise templates or parts could be produced whenever necessary. Since popular body shapes already exist, the expressions can be made using curve fits of points measured from existing instruments.

There is a problem with simple curve fits in that, no matter how the instrument is oriented with respect to the X and Y axes, some part of the curve has to be vertical and the resulting slope is infinite. One approach is to use cubic splines as implemented in most CAD software, but the result is not a simple expression.

An alternative is to transform the problem from rectangular to polar coordinates [89]. If the origin of the coordinate system is correctly chosen, a simple, closed-form expression can be used to describe the body shape. The origin of the coordinate system needs to be somewhere on the body so that no line drawn from the origin to the body side crosses any other point on the side or is nearly tangent to another point on the side. A point near the middle of the body usually works as shown in Figure 5.12.

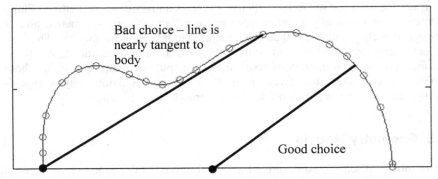

Figure 5.12 Choices for Origin of Coordinate System

Once the origin is placed, rectangular coordinates are turned into polar coordinates using two simple relationships

$$R = \sqrt{x^2 + y^2} \quad and \quad \theta = \tan^{-1}(x/y) \tag{5.31}$$

Where R is the radius and θ is the angle from the radius line to the horizontal.

Example 5.1 – A-Style Mandolin

An A-style mandolin has a pear-shaped body and it's relatively easy to find a good curve fit for it. The first step is to make a list of points in standard rectangular coordinates. Figure 5.13 shows rectangular coordinates measured from a representative instrument along with a plot verifying the results.

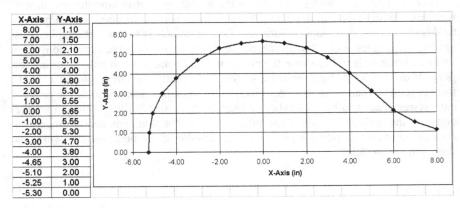

X-Axis	Y-Axis
8.00	1.10
7.00	1.50
6.00	2.10
5.00	3.10
4.00	4.00
3.00	4.80
2.00	5.30
1.00	5.55
0.00	5.65
-1.00	5.55
-2.00	5.30
-3.00	4.70
-4.00	3.80
-4.65	3.00
-5.10	2.00
-5.25	1.00
-5.30	0.00

Figure 5.13 Cartesian Coordinates for A-style Mandolin

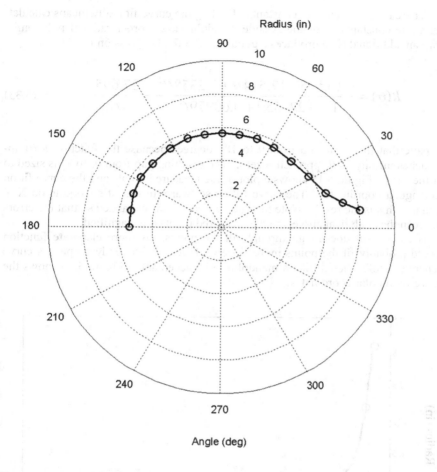

Figure 5.14 Polar Coordinate for A-Style Mandolin

The next step is to transform these 17 points into polar coordinates. The result is shown in Figure 5.14

There are many possible expressions that could be used to fit the measured data. Experience has shown that rational polynomials work well for many instrument body shapes.

$$R(\theta) = \frac{c_0 + c_2\theta + c_4\theta^2 + \cdots}{1 + c_1\theta + c_3\theta^2 + \cdots}$$

(5.32)

Where c_0, c_1, c_2, etc. are constants. Finding the curve fit really means calculating these constants. For the A-style mandolin, a 2/2 order rational polynomial with an additional term produces a good fit. The final expression is

$$R(\theta) = \frac{-16,505 + 17.514\theta + 3.2358\theta^2}{1 - 0.22455\theta + 0.60979\theta^2} - \frac{27,695}{\theta^2} \qquad (5.33)$$

Note that the angles go from 9 to 180 degrees because the function $R(\theta)$ approaches infinity as θ approaches zero. The resulting gap from 0° to 9° is sized to fit the neck. Figure 5.15 shows a plot of the measured points and the curve fit on rectangular coordinates. The curve fitting software 'thinks' it's working on X-Y data and this is the result. Note that the vertical axis is expanded so that the errors between the analytical curve and the measured points are magnified.

The curve fit doesn't go through all the points. A more elaborate function would probably fit the points more closely. However, the body shape this curve generates looks enough like a mandolin to be acceptable. Figure 5.16 shows the curve fit in polar coordinates.

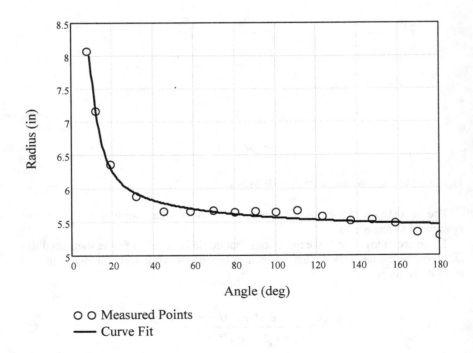

Figure 5.15 Body Curve Fit for A-Style Mandolin

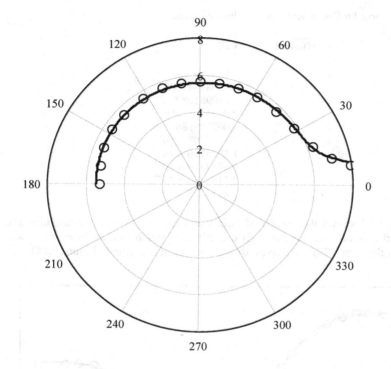

○ ○ Measured Points
—— Curve Fit

Figure 5.16 Curve Fit for A-Style Mandolin

Example 5.2 – Classical Guitar

The process used for the mandolin works for other instruments as well. The curve fit for a classical guitar is

$$R(\theta) = \frac{L}{25.5}\left[\frac{c_0 + c_2\theta + c_4\theta^2 + c_6\theta^3 + c_8\theta^4}{1 + c_1\theta + c_3\theta^2 + c_5\theta^3 + c_7\theta^4 + c_9\theta^5}\right] \qquad (5.34)$$

Where L is the scale length and the coefficients are as shown in Table 5.5. To scale the body up or down, simply insert the new scale length in for L.

Table 5.5 Curve Fit Coefficients for the Classical Guitar

Coefficient	Value
c_0	5.108849
c_1	−0.0089047088
c_2	−0.10981089
c_3	0.00012932766
c_4	0.002709986
c_5	2.4455108×10^{-6}
c_6	1.2823756×10^{-5}
c_7	2.3384339×10^{-7}
c_8	2.1193446×10^{-6}
c_9	$1.6911253 \times 10^{-10}$

Figure 5.17 shows the measured points and the curve fit on rectangular axes. As with the mandolin, the curve fit doesn't look much like an instrument when presented this way. On polar axes, though, it's clearly a guitar (Figure 5.18).

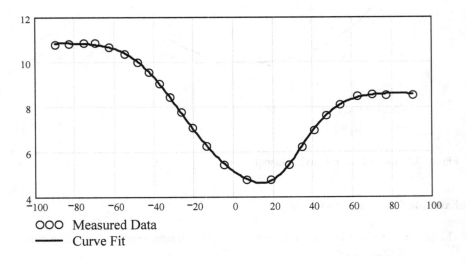

OOO   Measured Data
———   Curve Fit

Figure 5.17 Classical Guitar Curve Fit – Polar Data on Rectangular Axes

Figure 5.18 shows the curve fit in polar coordinates.

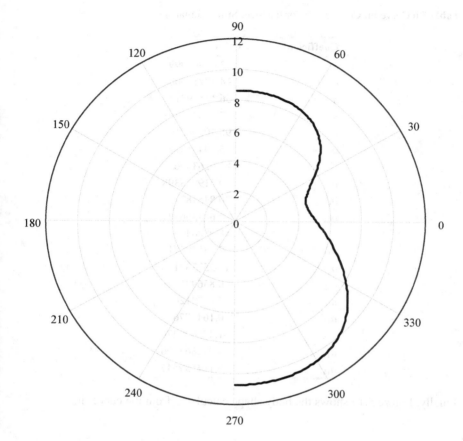

Figure 5.18 Classical Guitar Curve Fit on Polar Axes

Example 5.3 – Steel String Guitar

Finally, here's the same information for a steel-string acoustic guitar. Since the shape has sharper curves, a higher order rational polynomial is used. Even this more complicated function is compact compared to a long list of X-Y coordinates.

$$R(\theta) = \frac{L}{25.5}\left(\frac{a_0 + a_2\theta + a_4\theta^2 + \cdots + a_{16}\theta^8 + a_{18}\theta^9}{1 + a_1\theta + a_3\theta^2 + \cdots + a_{15}\theta^8 + a_{17}\theta^9}\right) \tag{5.35}$$

Table 5.6 presents all the parameters referenced in Equation (5.35)

Table 5.6 Curve Fit Coefficients for the Steel String Guitar

Coefficient	Value
a_0	12.69999999
a_1	−4.38217006
a_2	−56.1454981
a_3	7.545275491
a_4	98.70298019
a_5	−5.79491072
a_6	−81.9263396
a_7	0.519938398
a_8	20.84548718
a_9	2.644396721
a_{10}	18.9080405
a_{11}	−2.23520271
a_{12}	−19.3527871
a_{13}	0.856926696
a_{14}	7.652012607
a_{15}	−0.1648776
a_{16}	−1.47602927
a_{17}	0.012864596
a_{18}	0.114539547

Finally, Figure 5.19 shows the body shape calculated from the curve fit.

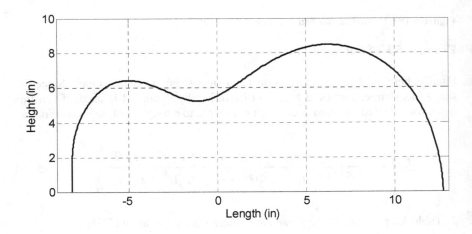

Figure 5.19 Steel String Guitar Curve Fit in Polar Coordinates

6 Manufacturing Processes

As the sales of new guitars in the US approaches 1 million units per year, manufacturing processes are clearly an important part of the engineering aspects of guitars. The best modern guitar factories are like the best factories in any other industry. They use computer-controlled equipment to perform many of the most precise tasks, greatly reducing the need for workers to be master craftsmen.

The differences in manufacturing processes used by small or individual builders and those producing on a factory scale are large enough to warrant separate sections here. However, advancing production technology has increasingly allowed small builders ready access to the same types of processes used in large factories; differences in the scale of production are not necessarily correlated with differences in production processes.

In order to characterize the different manufacturing processes and present them in some organized fashion, the distinction is made here between instruments made individually or in small batches and those made in large quantities in a traditional factory environment. Instruments made in small groups are those for which it is not practical to invest large sums in production tooling. Mass produced instruments are those for which the economies of scale justify complex, specialized equipment.

The size of production runs at different manufacturers probably forms a continuum. For the sake of organization, let's consider a small manufacturer to be one producing a few or, at most, a few dozen instruments per year.

6.1 Small Batch Instruments

The manufacturing processes required to produce small numbers of instruments can be very simple indeed. High quality instruments are routinely made by luthiers using hand tools. However, the use of simple fixtures and some well-chosen power tools can greatly speed the process.

A shop to make instruments one at a time or perhaps a few at a time can require relatively little capital investment. There is no standard configuration, but a productive space can be set up with a drill press, a band saw, a table saw and a belt/disc sander. The total investment would be a few thousand dollars at this writing. The addition of a thickness planer and a thickness sander would allow the luthier to buy wood in bulk and thin it to the desired dimensions as needed.

R.M. French, *Engineering the Guitar*,
DOI: 10.1007/978-0-387-74369-1_6, © Springer Science+Business Media, LLC 2009

Instruments made in small numbers often use extensive hand operations in the production process. Either the luthier is a hobbyist without the desire or resources to invest in a more sophisticated building method or is a small manufacturer who has determined that the return on investment does not warrant it. With minimal investment in dedicated tools and fixtures, operations of this size are often intended to produce unique custom instruments. The distinguishing features sometimes take the form of decorative elements like inlays. Figure 6.1 shows an elaborate inlay in the back of an acoustic guitar. The back itself is highly figured maple.

Figure 6.1 A Back Inlay by S. Balolia (image courtesy, S. Balolia, Grizzly Tools, www.grizzly.com)

Making instruments requires a succession of specialized assembly operations whose results must be relatively precise. Thus almost all luthiers (this one included) make a number of simple jigs to speed the process. These might include jigs to cut slots, glue the headstock joint and bend sides. Magazines and books on guitar making are a good source of clever designs for fixtures [90]. Figure 6.2 shows a simple fixture called a go-bar deck used to hold bracing while glue dries. Figure 6.3 shows a common side bending tool, a Fox universal side bender; this one is used at Taylor Guitars for making their high-end, low volume R. Taylor guitars.

Figure 6.2 A Simple Fixture for Installing Top Bracing

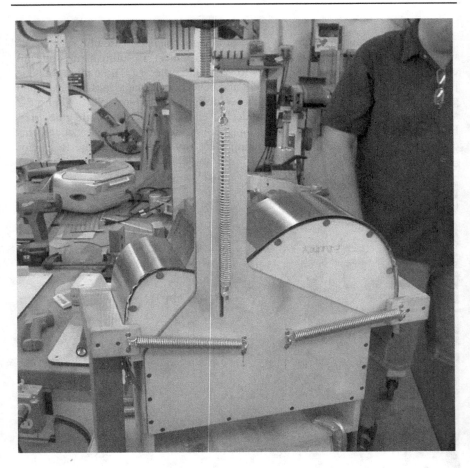

Figure 6.3 Simple Side Bending Jig (Picture by the author, reproduced courtesy, Taylor Guitars)

Simple fixtures like the ones shown here cost very little to make and are a great benefit in low production rate operations. When they are carefully designed, a skilled luthier can hold reasonably close tolerances while speeding production. If the most critical components of the fixtures are made using precision machine tools, the result can be an improvement in build quality and production rate.

Happily for the community of small luthiers, the cost of computer-controlled machine tools has decreased to the point that they are within reach of small manufacturers and hobbyists. Figure 6.4 shows a small computer controlled router cutting out a solid guitar body. The software required to lay out the instrument and drive the router are supplied with the machine and it is controlled by a simple PC. With a machine like this one, a small shop is capable of producing parts for a range of different instruments quickly and accurately.

Figure 6.4 A Small CNC Router Cutting Out a Solid Guitar Body

6.2 Mass Produced Instruments

The quality of mass produced instruments varies from the most basic instruments for beginners up to those suitable for concert musicians. As production is increasingly distributed around the world, the spectrum of mass produced instruments is broadening; good quality beginners' instruments are getting cheaper while the quality of high end instruments is improving even as their prices are often held constant.

This is likely due to a number of different circumstances. A major one, though, is the adoption of automated (computer-controlled) production processes by major manufacturers. Most major manufacturers are now making wide use of CNC machinery to both speed the production process and to greatly increase the precision of the resulting parts.

It is a basic tenet of quality production methods that build quality and build variation are inversely related. Being able to produce exactly the same part every time is the most fundamental of precursors to improving the quality of any mass produced product, especially guitars. It is important to note that this is true only of products that are intended to be identical, all having been built to the same design.

The characteristics that distinguish outstanding (and consequently more valuable) instruments from the merely competent ones are often quite subtle. Since small differences in nominally identical instruments – often too small to easily measure – can have a noticeable affect on the sound quality, production processes should be as precise and repeatable as possible given time and budgetary constraints.

Fortunately, precision equipment developed for other industries (and whose purchase price already reflects savings due to large sales volumes) are readily adapted to the task of making guitars. Figure 6.5 shows a standard industrial robot being used to polish guitar bodies after the finish (sprayed on by another robot) has cured. A single program is run for all guitar bodies of the same design, so there is great uniformity in the final finish.

The task of holding tolerances is particularly challenging in a large scale operation. A custom builder has the freedom to account for slight dimensional variations if necessary, but this is not practical in a factory environment. Current best practice is to precisely control the dimensions of parts at all stages of production. Figure 6.6 shows a rack of tops that have just been cut using a computer-controlled laser. The laser exerts no force that might distort or move the part and is generally accurate to within the precision of the mechanism that moves it (usually 0.001in or less).

Figure 6.5 An Industrial Robot Buffing a Guitar Body (image the author, reproduced courtesy Taylor Guitars, www.taylorguitars.com)

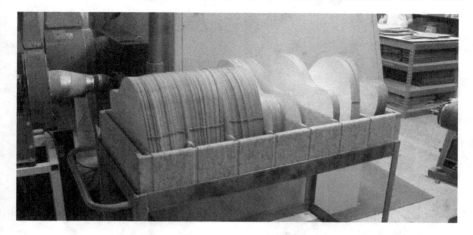

Figure 6.6 A Cart of Laser Cut Guitar Tops (Picture by the author, reproduced courtesy, Taylor Guitars)

Since the laser is computer controlled, features can easily be added to parts with simple modifications to the geometry model. For example, the tops shown in Figure 6.6 have a small triangular tab at the bottom as an alignment feature. They are visible as small dark areas that form a dark stripe as tops are laid next to one another. These and similar parts also have alignment marks produced by the same laser. The control computer simply adjusts the system so that only the top fibers are heated – like a very precise wood burner. The result is better alignment (i.e. higher precision) as the parts are assembled into a complete instrument.

For the beginning builder, few tasks are more challenging than bending the sides. Side benders of the type shown in Figure 6.3 are very popular, but are not well-suited to high volume production. A much-improved solution is the automated side bender shown in Figure 6.7. This device bends both sides at once and requires only that the straight blanks be loaded into the machine. The upper left image shows the bender after the lower heating element has been raised to form the waist and as the arms have begun to rotate up to bring the remaining straight portion of the sides in contact with the heating element. By the lower right image, the arms have rotated, pulling the sides up to final shape. Note the additional curvature in the lower bout, presumably to account for a slight springing back of the sides after being removed from the bender.

Another solution to the problem of accurately bending sides is to laminate sides from thin veneer, essentially producing plywood sides in the desired shape. Figure 6.8 shows side blanks made this way. The molded part is wide enough for several sets of sides to be simply sliced off the larger blank.

Figure 6.7 Automated Side Bender (images the author, reproduced courtesy Taylor Guitars, www.taylorguitars.com)

Figure 6.8 Laminated Plywood Sides (image the author, reproduced courtesy Taylor Guitars, www.taylorguitars.com)

One of the most important improvements in producing guitars in volume is the introduction of computer controlled milling machines. Such machines have been in use for decades in other industries with aerospace being one of the early adopters. Since first being developed, CNC milling machines have become ubiquitous in many industries. Accordingly, prices have decreased and the sophistication of both the hardware and software has increased.

In the absence of CNC machinery, parts are generally produced in volume by cutting tools following a pattern or template. Conceptually, there is a huge difference between following a pattern to make a part and making a part directly from the instructions stored on a computer disk. As an analogy, consider the difference between making a photocopy of a page produced on a laser printer and printing out another page. Just as the photocopied page will be less clear and might have blemishes that do not appear on the original, a part produced by following a CNC-produced template cannot be as accurate as one made directly using a CNC machine.

Another advantage of using computer-controlled equipment is flexibility in producing different parts on the same machines. Template-following cutting operations typically require precise setup since any misalignment of the template will be reflected in the final parts. Producing a different part requires installing a different template with the attendant alignment concerns. Producing a different part on a CNC tool often requires nothing more than running a different program. Thus, design changes can be easier to accommodate. It is also possible to produce a number of different designs on the same production line by simply switching programs as necessary. For example, the robot shown in Figure 6.5 is used to buff all instruments produced at that facility without any hardware changes. Figure 6.9 shows a group of eight necks being produced on a three-axis CNC milling machine.

Figure 6.9 A Group of Necks Being Machined (image the author, reproduced courtesy Taylor Guitars, www.taylorguitars.com)

6.3 Build Variation

A primary goal of volume production is to make all the parts exactly the same. It is commonly accepted in manufacturing operations that build variation is inversely related to build quality. Efforts to improve the product and the manufacturing process are hampered if the effect of changes cannot be distinguished from build variation. Thus, it is critical that good metrics for build variation are established. In guitar manufacturing, dimensions are routinely measured as part of the production process, but a more global metric would be valuable [91].

There are many possibilities for this, but the variation in the frequencies of lower modes is an attractive one. The structural-acoustic interaction through which an instrument makes sound is strongly conditioned by the coupled resonant frequencies of the instrument. Thus, controlling the variation in those frequencies is key to controlling the variation in the tonal quality.

To explore this idea, dynamic response measurements were made on a pool of instruments during the assembly process. Most of them had completed and finished bodies, but no necks. It was important that the body structure to be complete, including the finish. However, since the neck installation involves hand fitting, testing occurred 'upstream' of that process. If build variation was large for the bodies, it would be necessary to move to earlier stages of the build process to identify the source.

Taylor produces a range of different instruments in the same facility (see Figure 6.10) and all instruments available at that stage of assembly were tested. The largest group of a single design was the model 414, called a Grand Auditorium Cutaway. There are various build levels for this design, but the differences are limited to the side and back materials and complexity of the trim. All instruments had the same dimensions, the same bracing, the same bridge and the same soundboard.

6.3.1 Materials and Construction

One facet of guitar manufacturing that sets it apart from many other products is the variable nature of the raw materials. High quality guitars are generally made from solid wood (little or no plywood). Tops are sawn from logs selected to have straight grain and closely-spaced grain lines. Backs and sides are made from a wide variety of woods selected for their mechanical properties and appearance.

Tops are made almost exclusively from spruce, cedar and redwood. Traditionally, sides and backs are made from mahogany, rosewood or maple. However, the supplies of rosewood have been depleted and many manufacturers now use a wide range of tropical hardwoods such as ovangkol, sapele, cocobolo and bubinga. Thus, variation is introduced into the build process not only through the inherent variability of wood, but also through introducing different species with different mechanical properties.

Figure 6.10 A Collection of Guitars from the Taylor Product Line (image the author, reproduced courtesy Taylor Guitars, www.taylorguitars.com)

Since unfinished wood absorbs moisture readily from the atmosphere, wood which has already been air or kiln dried and then transported to the factory is generally allowed several weeks to reach equilibrium moisture content before being introduced into the building process. At Taylor Guitars, thin slabs are stacked with spacers as shown in Figure 6.11 and left in an open area before being sanded to the desired thickness. Blanks are then taken to a climate-controlled room where they are laser cut into tops and backs.

Most of the production process uses computer controlled equipment. There are still a few hand processes, though they largely involve decorative aspects of the instrument. Figure 6.12 shows the soundhole decoration being applied by hand. It should be noted, though, that the individual pieces are cut out using a computer-controlled laser and only the actual installation is done by hand.

Figure 6.11 Stacked Top and Back Blanks Being Conditioned Before Use (image the author, reproduced courtesy Taylor Guitars, www.taylorguitars.com)

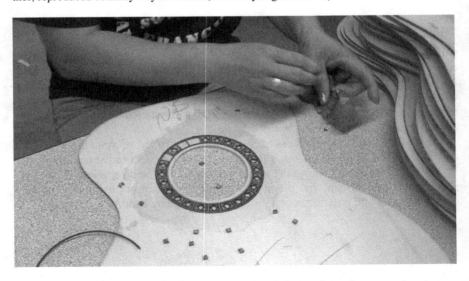

Figure 6.12 Soundhole Trim Being Applied by Hand (image the author, reproduced courtesy Taylor Guitars, www.taylorguitars.com)

Figure 6.13 Bodies and Necks after Finishing (image the author, reproduced courtesy Taylor Guitars, www.taylorguitars.com)

Figure 6.13 shows instruments after being finished, but before bridges are installed and the necks are fitted. Note that there are several different models on the cart and they are made from a range of different woods.

6.3.2 Response Variation Testing

A standard hammer impact test was performed on the instruments. The input source was a modal hammer with a hard plastic tip. The response was observed with a non-contacting laser displacement sensor (Keyence LK-G82). The sensor was placed so that the interrogation point was as close as possible to the hammer input point on the lower right side of the bridge. Figure 6.14 shows the test arrangement with hammer and laser sensor. Data was recorded with an Oros 4 channel data acquisition and the frequency response functions were calculated with a linear average of 5 taps. The bodies were supported by hard felt blocks placed around the edges so as not to influence the motion of the back.

A typical FRF and coherence function are shown in Figure 6.15. The resonant frequencies of the body are identified from the peaks in the FRF function. The coherence function shows how linearly related the input and output are. It serves, in this type of test, as a measure of the quality of the data. It should be noted that coherence is a necessary, but not sufficient condition for validity of the test data.

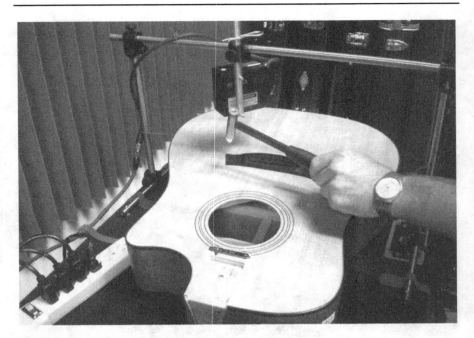

Figure 6.14 Guitar Body Being Tested (image the author, reproduced courtesy Taylor Guitars, www.taylorguitars.com)

Ideally, coherence is 100%, but it tends to drop at anti-resonant frequencies. This is because the response levels are very low at these frequencies and the signal-to-noise ratio is low. A higher proportion of noise necessarily reduces the correlation between the input and output. The result is that anti-resonance frequencies are more difficult to identify than resonance frequencies.

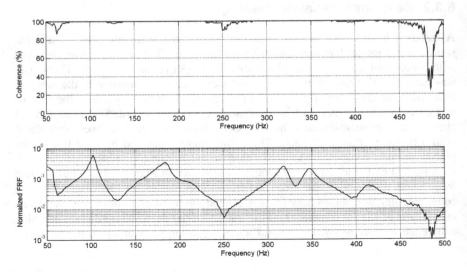

Figure 6.15 Typical FRF along with Coherence from Typical Impact Test

In order to ensure that the test procedure could distinguish test variation from build variation, data was recorded twice for most of the instruments so that isoplots^{TM} [92] could be constructed.

After collecting frequency response functions from the different instruments, the first two natural frequencies were identified along with the anti-resonance that fell between them. Figure 6.16 shows the first natural frequencies of the instruments sorted from lowest to highest. They are color/pattern coded according to the material used for the back and sides. In all cases, the tops were made from either Sitka spruce or Engelmann spruce, so it is assumed any part of the natural frequency variation due to differences in materials is due to the back and side materials rather than top materials. The mean frequency is 100.3 Hz and the standard deviation is 2.24 Hz (2.23% of the mean frequency).

It is also evident that the first resonant frequency is strongly correlated with the side and back material; sapele tends to result in lower frequencies and rosewood tends to result in higher frequencies. Some experienced luthiers believe that stiff sides improve the sound quality of guitars. It is interesting to note that rosewood has been the preferred wood for guitar sides since at least the late 1800s.

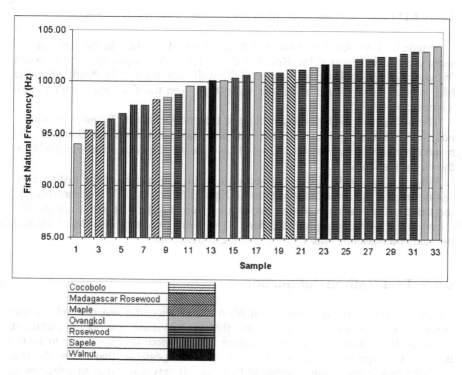

Figure 6.16 First Natural Frequencies of Instruments in Test Pool

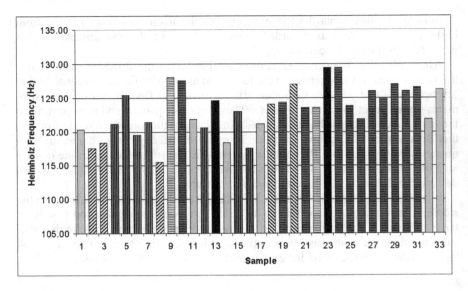

Figure 6.17 Helmholz Frequencies (Rigid Body Air Mode) of Test Instruments

Figure 6.17 shows the Helmholtz frequencies of the test instruments in the same order as in Figure 6.16. Recall that the Helmholtz frequency is the resonant frequency of the enclosed air volume if the wall were rigid; it is identified on the FRF plot as the anti-resonance (the valley) between the first two coupled resonance frequencies.

The 2-DOF and 3-DOF models suggest that the Helmholtz frequencies should not vary between instruments since the body geometry does not change. The mean Helmholtz frequency was 123.4 Hz and the standard deviation was 2.92% of the mean frequency.

Finally, Figure 6.18 shows the second natural frequencies of the instruments, again sorted in the same order as those in Figures 6.16 and 6.17. Not surprisingly, the trend of increasing resonant frequencies from Figure 6.16 (first natural frequencies) is generally reproduced here. The mean frequency was 180.1 Hz and the standard deviation was 3.11% of the mean frequency.

6.3.3 Test Method Verification

Intuition is often not sufficient to establish the validity of a test method – occasionally, one runs across test procedures that are 'obviously correct', but turn out not to be [93]. Whenever a test procedure is developed, it is important to verify that it gives repeatable results. If results are not repeatable, that means the test could be producing results dominated by error. It may also mean that the test is not measuring what it is intended to measure.

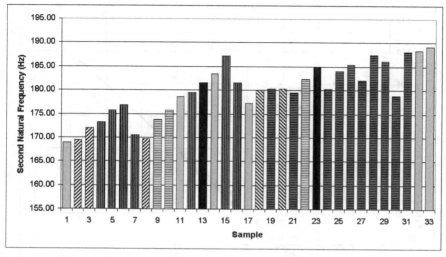

Figure 6.18 Second Natural Frequencies of Instruments in Test Pool

The isoplot™ method suggested by Dorian Shainin [92] is a simple tool to verify that the testing procedure can distinguish part variation from test variation. To construct an isoplot, the results of the second measurement are plotted against the results from the first. Distance between the points along the diagonal represents part variation. Distance perpendicular to the diagonal represents test variation. Thus, the ideal test would result in all data points along the diagonal. The general rule is that a box enclosing the data points should have an aspect ratio of 6 or more. If this is true, the test method is assumed to be capable of distinguishing test variation from part variation.

Since there was a relatively large pool of instruments, some of them were tested twice to establish statistically the repeatability of the method. Figure 6.19 shows the isoplot constructed from the first natural frequencies. The test variation is small compared to part variation, adding confidence that the trends suggested by the first natural frequencies are real and not an artifact of the test procedure.

Figure 6.20 shows that data from the second natural frequencies also satisfies the requirements of the isoplot method. Thus, the part variation is much larger than the test variation and trends suggested by this data can also be assumed to be real and not an artifact of the test procedure.

Finally, Figure 6.21 shows the isoplot made using the Helmholtz frequencies identified from the FRF plots. These frequencies are from the minima in the FRF plots between the first and second resonant peaks. In contrast to the encouraging results from the first and second natural frequencies, this data does not satisfy the requirements of the isoplot method; the test variation is not small compared to the part variation. This is not a complete surprise since the signal to noise ratio at an anti-resonance is, by definition, small. As is typical for this type of test, we often noted a significant decrease in coherence at anti-resonance frequencies. Thus, trends suggested by the Helmholtz frequency data should be evaluated carefully (perhaps with more testing) before being accepted as being accurate.

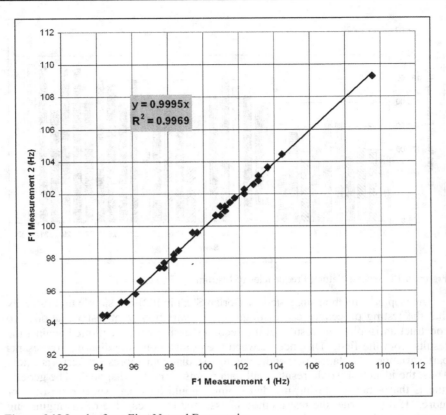

Figure 6.19 Isoplot from First Natural Frequencies

6.4 Building in Good Tone

Perhaps no goal is more elusive than that of building in sound quality. To date, objective measurements have not been conclusively correlated with subjective sound quality; there is, as yet, no number or even set of numbers that unambiguously predicts whether an instrument sounds good or not.

The difference between a good instrument and a superior one is subtle and can't always be distinguished just by listening. David Hosler, a product development manager at Taylor Guitars, said that one important measure is "how hard I have to work to make it sound good." The opinion of a skilled player must be combined with that of an experienced listener in order to rate an instrument with confidence. Fortunately, efforts to compare the subjective ratings of a group of instruments by skilled musicians have resulted in general agreement; musicians readily agree with each other when subjectively rating instruments [94,95]. Thus, we at least know that there is a 'right answer' even if we don't yet know how to measure or compute it.

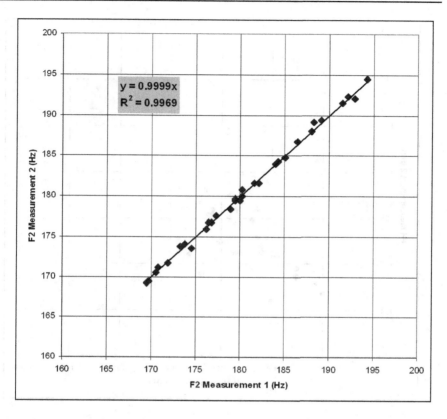

Figure 6.20 Isoplot from Second Natural Frequencies

The task is further compounded by the fact that luthiers of apparently equal eminence can often have directly opposing views about the same topic [45, 96]. Because of this widespread disagreement, it isn't possible to give a clear, unambiguous description of what is required to build good sound quality into an instrument.

If it is difficult to quantify the quality of an instrument, it may be harder still to quantify the skill of a luthier. Experience is necessary, but being prolific is, in itself, not sufficient. Numbers only matter if the individual instruments are generally better than those they succeed, and it is possible to make a long succession of equally mediocre instruments. A skilled luthier is capable of reliably producing superior instruments, and the most obvious metric is whether these instruments command high prices from skilled musicians.

Richard Bruné is a successful maker of classical guitars in Chicago. He is rare in that he was a professional musician before he became a luthier; while nearly all luthiers are capable of playing the instruments they make, relatively few are accomplished musicians. He began making guitars in 1966 while still in his teens and has made more than 675 during his career. He is a devoted collector of classical and flamenco guitars, believing that studying fine instruments is the best way to learn to make fine instruments [97].

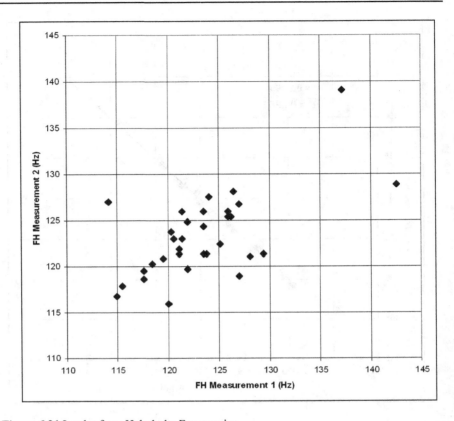

Figure 6.21 Isoplot from Helmholtz Frequencies

He routinely sells his instruments to skilled performers. Bruné has a bit of a trump card in his background, though, in that he has sold to two instruments to Andrés Segovia. In addition to being a universally acknowledged master of the classical guitar, Segovia was famously particular about selecting instruments. Selling him an insument was enough to secure the reputation more than one lu-thier. Thus, we can safely assume Mr. Bruné speaks with authority.

He has enumerated eight elements of classical guitar making he considers most important. These are, in order of importance:

1. Environment (temperature and humidity control)
2. Glue
3. Model or Design
4. Strings
5. Bridge
6. Neck Setup
7. Soundboard Design
8. Varnish

It is very interesting to note that choice of materials does not appear on the list. He states that great instruments can be made from less than ideal wood and offers the example of a violin made by Andrea Guarneri around 1680. It was sold at auction for nearly $200,000 in spite of the fact that it had a pronounced knot in the wood of the top between the f-holes. Interesting also is the fact that soundboard design appears next to last, suggesting that a range of different soundboard designs (with their attendant bracing patterns) can be used to produce fine instruments.

7 Sound Quality

When dealing with guitars, few topics are more controversial than sound quality. Skilled players and experienced listeners generally agree on subjective rankings of instruments, but the differences are notoriously difficult to measure and to describe using objective metrics.

Perhaps the fundamental problem is that descriptions of the sound of an instrument are almost universally imprecise. It is common for a guitar to be described as 'bright' or 'smooth' [98]. One particularly good sounding guitar that was louder than expected was described to me by an expert as 'a cannon'. The people offering these descriptive terms are clearly trying to find words whose emotional connotations are appropriate to his or her reaction to the sound. This type of allusion is the cornerstone of literature and poetry, but offers very little useful information to the analyst or the designer.

The first step in analyzing guitar sound quality is descriptive – the task of identifying the properties of the sound that are correlated with a superior instrument. In order to move closer to the underlying physical principles, the next task in the process is to identify the characteristics of the instrument that allow it to produce superior sound. The final task is to develop predictive (as opposed to descriptive) procedures so that superior sound quality is designed into the instrument from the start. For convenience, the three problems associated with guitar sound quality will be called:

- Description of sound quality based on subjective ratings
- Correlation of sound quality with physical features
- Prediction of sound quality from new design

Skilled luthiers regularly make high quality instruments, so these three problems have clearly been solved empirically. However, there is little in the technical literature to suggest that there is a firm technical foundation for any but the first one. The sound quality problem, already a difficult one, is compounded by the fact that different luthiers routinely offer differing and even contradictory conclusions regarding design features that produce superior sound quality.

One's first response might be to think that they can't all be right at once. However, maybe they could be. There is no single definition of good sound quality for the obvious reason that different players prefer different tonal qualities in their instruments. Music is, after all, an art and diversity of expressions is part of the point. Additionally, it is also far from clear that only a single combination of

R.M. French, *Engineering the Guitar*,
DOI: 10.1007/978-0-387-74369-1_7, © Springer Science+Business Media, LLC 2009

design characteristics can produce a specific type of sound. It is certainly possible that many different combinations of instrument characteristics such as top stiffness distribution, body shape, material properties, etc. can result in a pleasing instrument.

The differences in acceptable tonal qualities tend to be subtle and the measurable differences between competent and superior instruments can be very small. Fortunately, there is broad agreement on what constitutes poor sound quality. Furthermore, there is also broad agreement on design elements that produce inferior instruments. The result is that there are general guidelines that will reliably produce a good, though perhaps not superior, instrument. Let us start with the first element of sound quality, objective descriptions of what sounds good and what doesn't.

7.1 Elements of Sound Quality

While different players tend to prefer different nuances in the tone produced by their instruments, there are some characteristics of the sound produced by guitars that are generally preferred. These characteristic are sometimes best described in time domain, sometimes in frequency domain and sometimes in time-frequency domain. Let us begin with the time domain descriptions.

7.1.1 Time Domain Descriptions

When a string is plucked and allowed to ring, there are two different components in the resulting motion: attack and decay. Figure 7.1a shows about the first second of the response of a low E string. The attack portion of the response occurs in about the first 40 ms and the remainder is the decay. This data was recorded in a large hemi-anechoic chamber with a background level of approximately 28 dB(A). Essentially all the background noise in this chamber is below 200 Hz, so a 70 Hz high pass filter (12th order Butterworth) was applied to the raw data. Figure 7.1b shows the first 200 ms of the signal. The signal clearly builds over the time from 0.02 s to 0.05 s, a span of 0.03 s (30 ms). In Figure 7.1b, the trace shows a high frequency oscillations superimposed over the low frequency signal. This is due to the fact that there are many frequencies present at once. In order to make the attack and decay easier to see, the time signal can be processed further. Figure 7.1c shows the absolute value of the original time signal after a low pass filter has been applied. This is not, perhaps a standard analysis and the resulting amplitude doesn't match that of the original signal. However, the attack and decay are clear.

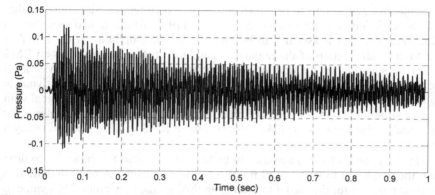

Figure 7.1a Response of the Low E String in Time Domain

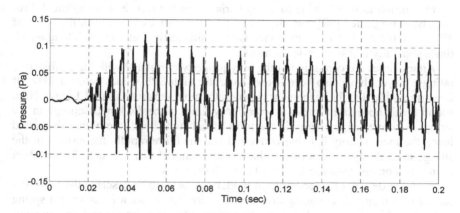

Figure 7.1b Attack and Beginning of Decay of Low E String

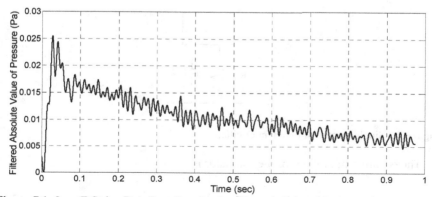

Figure 7.1c Low E String Data Post-Processed to Show Attack and Decay

Subjective evaluations have shown correlations between superior sound quality and low variation of attack times between strings [99]. Attack time doesn't have a fixed definition, but is assumed to be the time for the amplitude of the signal to rise from small fraction of the maximum amplitude (often expressed as an RMS value) to some large fraction of the maximum amplitude. The RMS level oscillates as long as the averaging time is not too long (the standard for dynamic signals is 135 ms). Thus, it makes sense to fit the rising portion of the signal with an appropriate function. A good definition for attack time is the time required for the signal to rise from 10 to 90% of the maximum RMS value as determined by a linear fit [94].

Similarly, decay time is assumed to be the time required for the signal to drop from the maximum to some small fraction of the maximum. A good definition is the time required for the signal to drop from 90% of the maximum RMS value to 10% of the maximum RMS value as determined from an exponential curve fit.

In addition to small string to string variation in the rise time, experimental results have suggested [99] that players prefer longer attack times – on the order of 40 ms. In the pool of 13 instruments used for this testing, guitars with long attack times generally had shorter decay times.

Attack times are certainly not the only property correlated with high subjective ratings; players have also been shown to prefer high SPL values below 1 kHz. It is worth noting that the human ear is generally most sensitive around 1 kHz. High sound levels necessarily mean shorter decay times since the kinetic energy in the strings is converted to energy in the form of radiated sound; the higher the sound level, the more energy is being converted and the more rapidly the motion of the strings must decay. It can help to have a math model that accounts for the short term time domain response in at least a qualitative way.

A single degree of freedom math model can be used to describe the basic behavior of a damped, vibrating structure. Figure 7.2 shows a mass with a spring and a damping element. Mass is m, the spring stiffness is k and the damping coefficient is c.

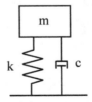

Figure 7.2 Single Degree of Freedom Oscillator

The equation of motion for the oscillator is

$$m\ddot{x} + c\dot{x} + kx = f(t) \tag{7.1}$$

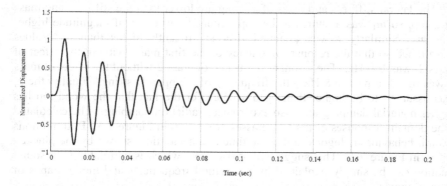

Figure 7.3 Exponential Decay of Single DOF Oscillator

The mass' response to an impulsive force is shown in Figure 7.3. The equation of motion assumes that damping is proportional to velocity; this is by far the most widely assumed damping mechanism. The position of the mass increases during the first 0.01 s, which was the duration of the input force, and settles into a sinusoidal oscillation with an exponential decay. There are other damping models in the literature, but this one works well for a range of applications.

In a guitar, damping is largely due to energy loss from two sources, material damping and sound radiation. Material damping generally produces exponential decay. Acoustic radiation is a direct function of the velocity of the moving surface, so it is reasonable for radiation damping to be modeled as proportional to velocity as well.

The single degree of freedom response shows the effect of proportional damping, but not of two interacting flexible structures (like a top and a string). Qualitatively, this interaction can be modeled using a 2-DOF system as shown in Figure 7.4. Clearly, this model is far too simple to capture the physics of a guitar, but it does clearly display dynamic interaction similar to that seen from a string mounted to a flexible body.

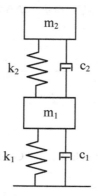

Figure 7.4 Two Degree of Freedom Oscillator

The lateral stiffness and mass of a string are lower than the stiffness and mass of a top, so m_1 was assumed to be approximately an order of magnitude higher than m_2. Similarly, k_1 was assumed to be significantly higher than k_2. Values were tuned so that the resonant frequency of the first mass, without the effect of mass 2, was 100 Hz. The resonant frequency of mass 2 without the effect of mass 1 was tuned to be 146.8 Hz – the frequency of the open D string. Finally, the c_1 was assumed to be higher than c_2 to account for the fact that the wood top has higher material damping and also experiences radiation damping. The calculated time domain responses of the two masses are shown in Figure 7.5. Both motions exhibit behavior analogous to the rise time and decay time shown by the plucked string in Figure 7.1. This suggests that rise time, which is correlated with sound quality, can be simply explained by the natural frequencies and mode shapes of the instrument.

7.1.2 Frequency Domain Descriptions

The tone produced by a guitar is strongly conditioned by the frequencies that are produced and their relative magnitudes. Figure 7.6 shows a spectrogram of the sound produced by a dreadnaught guitar being strummed with open strings. The signal shown in Figure 7.1 is simply the first second of this longer recording. For clarity, Figure 7.6 shows only the components below 1800 Hz – there are measurable components all the way up to 20 kHz. It is clear that there are many harmonics present for each of the six strings. Additionally, there are frequencies at which there are several closely-spaced string resonances. For example, at approximately 1000 Hz, there are at least five closely-spaced string frequencies.

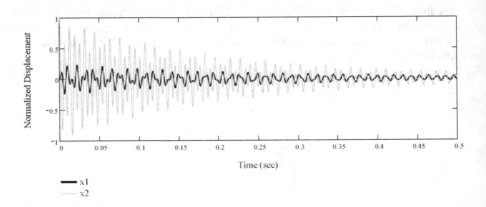

Figure 7.5 Response of 2-DOF Oscillator to Pluck Initial Condition

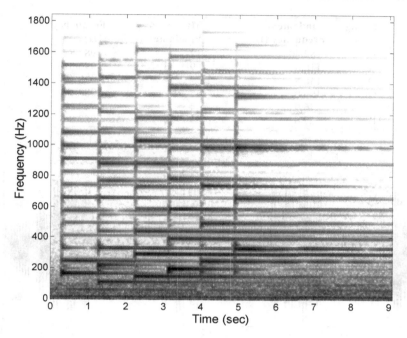

Figure 7.6 Spectrogram of Open String Strum on Taylor 710

It is well known that sinusoidal waves with closely spaced frequencies will produce a beat pattern in which the envelope of the combined signal is periodic and whose frequency is related to the difference in the frequencies [17]. Table 7.1 shows the open string frequencies of the six strings and the harmonics close to 1000 Hz. Figure 7.7 shows a synthesized signal made by simply combining, in equal proportion, sine waves of the frequencies shown in Table 7.1. Finally, Figure 7.8 shows a close-up of the spectrogram centered around 1000 Hz. After the 6th string is plucked, there is clearly an amplitude variation consistent with that shown in the synthesized signal.

It is possible to identify fundamental frequencies and a large number of harmonics from recordings of the radiated sound produced by instruments of just about any subjective sound quality. Clearly, the link between objective descriptions such as frequency content and subjective sound quality ratings is subtle. First order descriptions are not sufficient and it seems quite likely that less obvious effects such as the interaction of different spectral components, like the beat effect mentioned above, are important.

Table 7.1 Harmonics Closest to 1000 Hz

String	Fundamental Frequency (Hz)	Multiple of Fundamental	Frequency (Hz)
E	82.40	12	988.8
A	110.0	9	990.0
D	146.8	7	1028
G	196.0	5	980
B	246.9	4	987.6
E	329.6	3	988.8

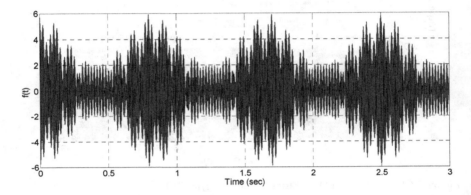

Figure 7.7 Beat Synthesized from String Frequencies

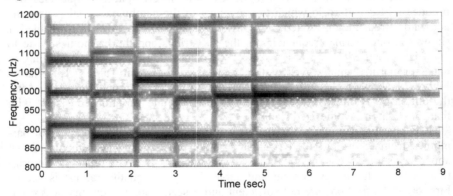

Figure 7.8 Beat Phenomenon in Measured Acoustic Response

Finally, Figure 7.9 shows a waterfall plot of the open string strum, essentially a different presentation of the plot in Figure 7.2. that is easier for some readers to interpret. It is clear that the acoustic response is completely dominated by the frequencies due to the vibrating strings. There are no peaks due to coupled acoustic-

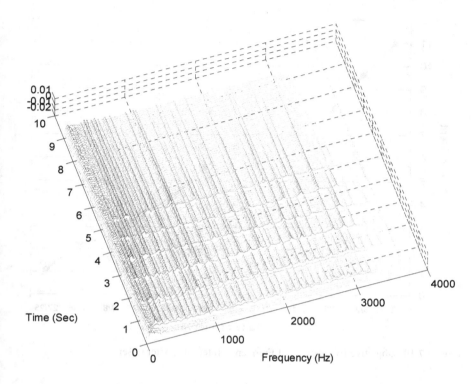

Figure 7.9 Waterfall Plot of Spectrogram from Open String Strum

structural response of the body. The conclusion is that the sound produced by an acoustic guitar, either a good one or a bad one, is dominated by the string vibrations and that the dynamic response of the body serves mostly to change the proportions of each of the string frequencies that are present in the radiated sound.

One proposal for describing the quality of the sound from acoustic guitars is based on the concept of consonance and dissonance [30]. Two or more sounds that are pleasing to the human ear are termed consonant and ones that are not are termed dissonant. This simple pair of definitions touches on the problem of defining sound quality; there is no absolute definition of what combinations of sounds people tend to like. The rule of consonance-dissonance (RC-D), proposed by Šali and Kopač [100], is based on relationships between subjective ratings and frequency intervals. One of these is shown in Figure 7.10. The degree to which a specific frequency ratio is pleasing is called, in the original reference, the order of merit and they are shown here in order of increasing dissonance (decreasing consonance). The RC-D method develops a numerical rating of sound quality based on sums of the magnitudes of frequency components.

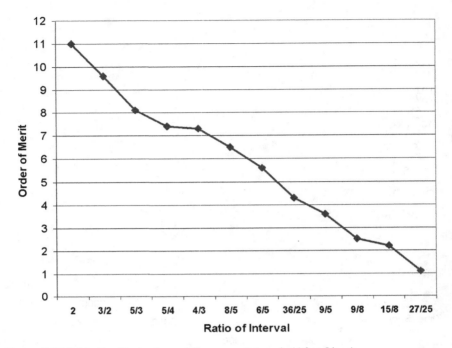

Figure 7.10 Subjective Evaluations of Frequency Intervals (After Olsen)

Šali and Kopač also note that guitars exhibiting superior sound quality have higher peaks in the FRF at the first and second body resonances, accompanied by lower damping. This supports the idea that simple models with only a few degrees of freedom are useful in analyzing coupled dynamics of guitars.

7.1.3 Psychoacoustic Descriptions

The field of psychoacoustics is devoted to the study of subjective human perceptions of sound; the subjective, emotional response to music is clearly within its realm. Human perception of sound is quirky and includes some characteristics that are far from intuitive. The psychoacoustic of timbre are of particular importance when studying musical instruments.

Perhaps the most important concept in applying psychoacoustics to musical instruments is that of critical bandwidth. The human ear is not capable of distinguishing individual frequency components of a sound if those components are very close in frequency [30]. If pure tones of two frequencies, F_1 and F_2, are played together, perception of the sound depends on the frequency spacing. If the two frequencies are identical, then the notes sound in unison. If the difference is less than about 13 Hz, the result is a beat sound. The beat frequency is simply the frequency difference between the two tones, $F_1 - F_2$. As the frequency difference increases, the beat sound is replaced by a composite tone with a rough character. When the frequency difference is increased still further, the rough sensation subsides and the result is a sound consisting of two smooth tones. The critical bandwidth is simply the frequency difference that separates the sensation of two rough tones from two smooth ones.

The boundary that defines the critical bandwidth is not a hard one and varies between listeners. In fact, it has been shown to have its basis in how the inner ear reacts to sounds of different frequencies. An extremely important feature of the critical bandwidth is that it varies with frequency [101]. An approximate expression, valid in the range of 100–10,000 Hz, for critical bandwidths is

$$CB = 6.23 \times 10^{-6} f_c^2 + 93.39 \times 10^{-3} f_c + 28.52 \qquad (7.2)$$

Where f_c is the center frequency in Hz. Thus, at 200 Hz, the critical bandwidth is 47.5 Hz and at 2000 Hz it is 240 Hz.

The concept of critical bands is important when considering the harmonics of a note. Harmonic frequencies increase linearly, but the width of the critical band increases as the square of frequency. Thus, there should be some frequency beyond which the average listener could not distinguish individual frequencies. In general, this happens around the 7th harmonic. For example, the fundamental frequency of the A string is 110 Hz and the frequency of its 7th harmonic is 880 Hz (110 Hz is the fundamental frequency, 220 Hz is the first harmonic frequency and so on). At 880 Hz, the critical bandwidth is 116 Hz. Thus, the average listener could not distinguish the 7th harmonic from the 8th. As an additional example, consider the high E, whose fundamental frequency is 329.6 Hz. The frequency of the 7th harmonic is 2637 Hz – well within the human hearing range – and the critical band is 318 Hz. This does not suggest that higher harmonics are not important in creating the timbre of musical instruments, only that our perception of them changes as the critical band width exceeds the fundamental frequency. Figure 7.11 shows the critical band curve along with fundamental frequencies of the A and high E strings.

A byproduct of our inability to perfectly distinguish closely spaced frequencies is auditory masking. The details are beyond the scope of this discussion, but the basic idea is simple. When two or more tones are heard together under the right circumstances, the lower frequency tone can mask the higher one. In psychoacoustics literature, the effect is usually described in terms of pure tones, but masking also occurs between individual frequency components of a complex sound. The masking effect is a function of both frequency and amplitude, but is shown in qualitative terms in Figure 7.12. The component of the signal that remains apparent is called the masker.

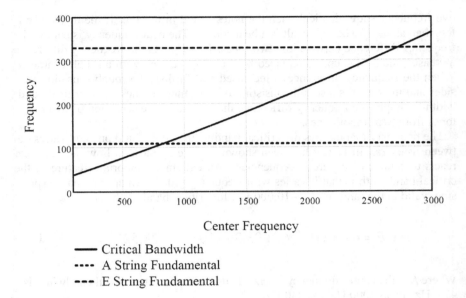

Figure 7.11 Critical Band Frequencies with A and High E String Frequencies

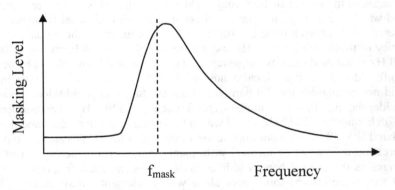

Figure 7.12 Masking Level as a Function of Frequency

Frequency components that are completely masked don't contribute to the perceived sound; they are still present and can be seen in spectrograms of the recorded sound, but human listener is unlikely to notice their presence. By extension, not all of the sound components produced by a guitar can necessarily be distinguished by a human listener. In fact, masking is one of the phenomena exploited to compress sound files using schemes like MP3 [102].

Another quirk of human hearing is that we don't always perceive the pitch of a note correctly. In one experiment, subjects presented with a pure tone of a given frequency, f_1, were asked to adjust the frequency of a second tone, f_2, so that it was an octave higher than the first. As long as f_2 was less than 2500 Hz, test subjects were generally able to find the point where $f_2 = 2 f_1$. However, as the lower frequency was raised above 2500 Hz, the results deteriorated. It appears that people generally find it difficult to perceive octave intervals unless both frequencies are below 5 kHz [103]

7.1.4 Subjective Rankings

Guitar sound quality ratings are primarily subjective, so there is a body of work in the technical literature describing efforts to rate instruments using groups of experts who listen to and play them. It is interesting that, when a group of musicians or trained listeners evaluates a group of instruments, they tend to agree on the order in which the instruments should be ranked according to sound quality. Subjective evaluations of groups of instruments typically use the method of paired comparisons [104], a technique in which a succession of comparisons between two instruments leads to a ranking of the entire group of instruments.

The number of paired comparisons that can be made between n different instruments can be shown graphically at least two different ways as shown in Figure 7.13. The number of possible paired comparisons is related to triangular numbers as shown. In algebraic terms, the number of possible comparisons is

$$N = \frac{n^2 - n}{2}. \tag{7.3}$$

Since the nature of the evaluation is subjective, it is not uncommon to use subjective descriptors such as bright or smooth in place of more clearly defined metrics. Typically, the participants are asked to select one or the other of a pair of opposing adjectives such as Clear – Diffuse and Sharp – Smooth.

Once subjective rankings have been established, objective measurement are then compared in an effort to find a correlation. The following characteristics have been correlated with good sound quality

- Long attack time
- Low variation in attack time between strings
- High sound level below 1000 Hz
- High amplitude and low damping in FRF of first two resonant frequencies
- Adjectives Clear, Sharp and Brilliant

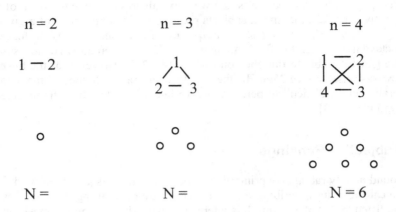

Figure 7.13 Possible Numbers of Paired Comparisons

7.2 Structural Characteristics

Once the basic elements of good tonal quality are described, the next step is to identify the structural features that are important in producing the tonal quality. Much has also been written about tap testing [105] and the selection of appropriate materials [106]. Additionally, many articles have appeared in journals devoted to lutherie in which methods are proposed for modifying the structure (usually the braces) to improve sound quality [107].

The process for selecting wood is fairly universal, though it would appear to be built on a fairly tenuous technical footing. It is standard practice to flex a piece of wood manually to check cross-grain stiffness as shown in Figure 7.14.

Because wood is formed from strong fibers joined by a weaker material, it is structurally very similar to a fiber composite in which the fibers all run the same direction. For obvious reasons, this is usually called a unidirectional composite. Since the fibers are all parallel (in either wood or our notional composite), the stiffness along the grain is different than that across the grain.

Metals and un-reinforced plastics have the same properties everywhere and the same stiffness in any direction. In strength of materials texts [40, 58], such materials are called homogenous and isotropic. In wood, the two axes in which stiffness must be defined are along the grain and across the grain. Materials with this property are termed orthotropic, presumably because the two directions in which stiffness is specified are orthogonal to one another. In matrix form, the 3-D Hooke's law for orthotropic materials is:

Figure 7.14 Checking a Piece of Wood by Flexing Across the Grain

$$
\begin{Bmatrix}
\varepsilon_x \\
\varepsilon_y \\
\varepsilon_z \\
\gamma_{xy} \\
\gamma_{yz} \\
\gamma_{zx}
\end{Bmatrix}
=
\begin{bmatrix}
\dfrac{1}{E_x} & -\dfrac{\nu_{yx}}{E_y} & -\dfrac{\nu_{zx}}{E_z} & 0 & 0 & 0 \\[2ex]
-\dfrac{\nu_{xy}}{E_x} & \dfrac{1}{E_y} & -\dfrac{\nu_{zy}}{E_z} & 0 & 0 & 0 \\[2ex]
-\dfrac{\nu_{xz}}{E_x} & -\dfrac{\nu_{yz}}{E_y} & \dfrac{1}{E_z} & 0 & 0 & 0 \\[2ex]
0 & 0 & 0 & \dfrac{1}{G_{xy}} & 0 & 0 \\[2ex]
0 & 0 & 0 & 0 & \dfrac{1}{G_{yz}} & 0 \\[2ex]
0 & 0 & 0 & 0 & 0 & \dfrac{1}{G_{zx}}
\end{bmatrix}
\begin{Bmatrix}
\sigma_x \\
\sigma_y \\
\sigma_z \\
\tau_{xy} \\
\tau_{yz} \\
\tau_{zx}
\end{Bmatrix}
\tag{7.4}
$$

In the 3-D formulation, the X, Y and Z directions might correspond to the direction parallel to the fibers, parallel to the growth rings and the direction perpendicular to the growth rings. A thin plate like a top or back is essentially two-dimensional; the 2-D form of the equation is created by simply eliminating the z terms.

$$
\begin{Bmatrix} \varepsilon_x \\ \varepsilon_y \\ \gamma_{xy} \end{Bmatrix} = \begin{bmatrix} \dfrac{1}{E_x} & -\dfrac{v_{yx}}{E_y} & 0 \\[2mm] -\dfrac{v_{xy}}{E_x} & \dfrac{1}{E_y} & 0 \\[2mm] 0 & 0 & \dfrac{1}{G_{xy}} \end{bmatrix} \begin{Bmatrix} \sigma_x \\ \sigma_y \\ \tau_{xy} \end{Bmatrix} \tag{7.5}
$$

If the grain direction in a plate is called the X direction and the cross-grain direction is called the Y-direction, the material properties needed to describe the stiffness are E_x, E_y, and G_{xy}. However, the cross grain bending test only determines (very approximately) E_y. It is implicit in this rough test that desirable material properties are correlated with high cross-grain stiffness.

It is also common to hold a plate up to the ear, gripping it loosely between two fingertips and tapping it. Preferred materials appear to have high cross-grain stiffness and low damping. Recall that low damping in the first two body modes is correlated with good sound quality. Preferred species for soundboards are Sitka or Englemann spruce, western red cedar or, less frequently, redwood. They are distinguished from other species of close-grained softwoods by a large stiffness to mass ratio.

In broad terms, the task of making a guitar top is that of making a light structure that is strong enough to resist string tension while simultaneously being flexible enough to respond to the dynamic string forces. The most common way of trying to achieve this goal is with different bracing patterns. The most common ones have been addressed earlier. No matter what the bracing pattern is, the basic mechanics of the beam-like braces themselves are unchanged. The stiffness of a brace is proportional to the cube of its height and proportional to its width. This doubling the width of a brace doubles its weight and its stiffness. Doubling its height doubles its weight, but increases the stiffness by a factor of 8. Thus, braces that are relatively narrow and tall will yield a lighter structure for a given weight.

It is also desirable for the stiffness of the top to be distributed so there are no sharp local changes in slope or curvature anywhere. These can cause the top to split or crack. Thus, it is unusual for a flat top guitar to have a few large braces; it is much more common for a larger number of smaller braces to be used (the structural characteristics of archtop instruments are different and these instruments generally use two large braces and higher top thickness). While most braces are made of single pieces of wood, there have been successful experiments with braces made of light wood with thin strips of graphite bonded to the upper surface [108]. The idea of laminated braces is not new; the Maurer Company made instruments with a vertical strip of rosewood sandwiched between strips of spruce.

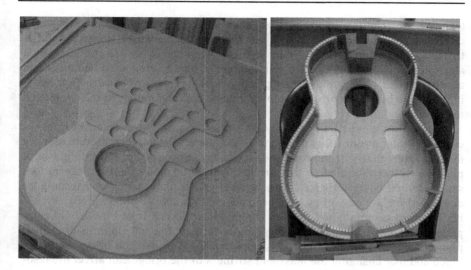

Figure 7.15 Three-Piece Top Showing Spacer and Stiffener

One unconventional experiment (this one by the author) is to use a thin sheet of aircraft-grade plywood to add stiffness over a large portion of the top with a machined spacer made of light ¼ inch thick plywood as shown in Figure 7.15. This approach has the advantage of being very simple since the top has only three pieces – the top plate, the spacer and the plywood stiffening sheet. Furthermore, all three components are easily cut out on a CNC router, a laser cutter or a router following a simple template. The design of the spacer can also be varied to change the stiffness distribution as desired. Finally, the materials are inexpensive and very uniform, so a good design can be reproduced easily, relatively cheaply and with minimum build variation.

The instrument shown in Figure 7.15 is a proof of concept prototype. The sound quality is good and can probably be further improved with modifications to the spacer plate. It is also durable since plywood is generally less affected by humidity changes and there is a large gluing surface between the spacer and the top. Furthermore, assembly time and material costs were low.

The most sophisticated efforts to correlate physical features with sound quality tend to work on some metric based on dynamic structural response. Attempts to correlate specific combinations or ratios of body resonant frequencies with sound quality have produced ambiguous results. In the violin world, much has been written about plate mode tuning [109], but there is still debate over whether it represents a repeatable process or relies on the intuition of the luthier. It seems likely now that a robust objective description of the structural characteristics that improve sound quality will be more sophisticated than frequency ratios.

Many luthiers tap test their instruments during construction. This practice is sometimes called voicing and has a long history, but is still the source of some debate. In the interest of full disclosure, the author admits to being agnostic on the subject; many skilled and accomplished luthiers depend on it, but robust technical descriptions of the process need to be developed. Tap testing may consist of the builder simply tapping on the instruments with a fingertip and listening to the resulting sound. More sophisticated approaches are not uncommon and some builders even use lab test equipment like instrumented hammers and accelerometers. It is common to tap test at various stages of construction, starting with the blanks from which tops are made and continuing to the completed instrument.

Tap testing is essentially a method of measuring the lower resonant frequencies of a guitar or one of its components and the results are then used to identify modifications that will improve sound quality. Specifically, the top and back plates are often modified so that the frequencies or mode shapes conform to a set of desired characteristics. This is often called plate tuning and techniques developed for use on violins have been adapted for use on guitars [110]. Probably the most common modification is removing material from braces. On completed instruments, this is done by going in through the soundhole with a small plane or, less often, a chisel. It is important to note that the bridge acts as a brace and that changing the mass and stiffness of the bridge can affect the resonant frequencies of the top.

There are two basic assumptions underlying tap testing during construction. The first is that guitar sound quality is correlated, at least subjectively, with the distribution of resonant frequencies and their amplitudes. The second is that frequencies of components (usually the top and back) determined before the instrument is assembled can be used to predict sound quality after the instrument is assembled.

The first assumption, that certain combinations of frequencies are correlated with good sound quality, is asserted by many articles in the technical literature. One commonly stated goal is that the first resonant frequencies of the top and back should be close to each other [111]. The goal is to create a second body mode from the two closely spaced modes coupled by the enclosed air. A closely related design goal is for the first mode of the instrument to be in the range of 90 Hz – 100 Hz (for full-sized instruments). Satisfying these conditions should produce an instrument that responds well at low frequencies – the fundamental frequencies or lower harmonics of the strings.

In order to measure the resonant frequencies of the top and back separately, one must either measure the frequencies of the free plate and predict the frequencies after the plates has been glued to the sides or create a test fixture into which the top and back can be mounted for testing. The first option requires a sophisticated predictive method like finite element analysis. Lacking this, accurate results can be obtained with an appropriately designed fixture. Figure 7.16 shows a test fixture into which tops and backs can be easily mounted. They can be tested individually and with the correct boundary conditions, but without the effect of the enclosed air volume. Using toggle clamps to hold the top and back in the fixture also allows them to be removed easily and replaced after being modified.

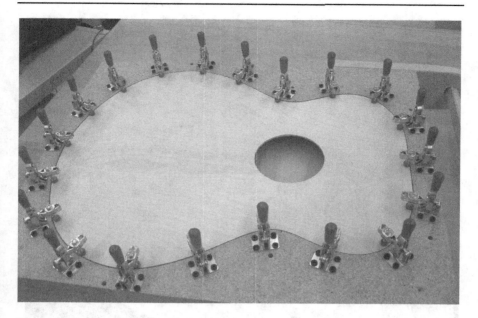

Figure 7.16 A Test Fixture for Tops and Backs

The second assumption, that the frequencies of the top and back before the instrument is assembled can be used to predict sound quality is, from an engineering standpoint, not obvious. Changing the boundary conditions of even a simple plate can significantly change both the resonant frequencies and their corresponding mode shapes. On a structure as complex as a braced guitar soundboard, the change is potentially greater and harder to predict.

One goal of tap testing is to identify portions of the soundboard that do not respond dynamically and, thus, do not radiate sound. Increasing the radiating area of the soundboard can increase the volume of the instrument and may affect tone as well. One structural modification that can be explained using basic physical principles is thinning the top or back plate near the edge. This has the effect of reducing the bending stiffness near the edges of the top and giving the top a more nearly pinned boundary condition. The result is that a larger portion of the soundboard can be in motion due to string vibration and more of the kinetic energy can be radiated as sound. With this intention, at least one manufacturer routs a groove near the edge of the lower bout of tops as shown in Figure 7.17

Finally, some mention needs to be made of the effects of the finish on sound quality. Finish protects the wood, but adds mass along with its own stiffness and damping properties. Less obviously, it may also affect the radiation by changing the acoustic absorption – a measure of how air interacts with the moving surface of the instrument. However, a body of compelling data describing this effect appears to be missing from the technical literature.

Figure 7.17 Groove Around Soundboard Edge to Reduce Bending Stiffness, Patent 6,759,581. (image the author, reproduced courtesy of Taylor Guitars, www.taylorguitars.com)

The idea of finish affecting sound quality probably originated in violin making where there is a strong tradition of developing proprietary finishes. These were originally oil-based varnishes, though spirit varnishes (alcohol-based) eventually came into use. The literature on violin making is liberally sprinkled with discussions of varnish formulations and their effects on the finished instruments. It is also not uncommon for makers to guard their varnish formulations as one would a family recipe.

Traditions among guitar makers are less established. Luthiers in Spain in the 19th century generally used shellac and it is sometimes still used for classical guitars. In the United States, manufacturers of steel-stringed guitars adopted nitrocellulose lacquer which was then the common industrial finish. Manufacturers now generally use modern industrial finishes. Lacquer has been almost completely superseded and is now generally available as a specialty item.

There appears to be no general agreement on the effect of finish on the sound of acoustic guitars, though many luthiers believe that a thin finish is preferable. Finish thickness is carefully controlled in most factories, a task made easier by computer-controlled equipment.

7.3 Acoustic Characteristics

Clearly, the sound quality of guitars is dependent on their ability to radiate sound. Section 2.4 touched on the basics of radiation from a moving surface; now, the discussion turns to the specific radiation characteristics of a guitar [112]. To start, it is useful to have a conceptual model of how the different components radiate sound. Figure 7.18 shows high and low frequency radiation from the guitar in schematic form.

Any sound energy radiated by an unamplified acoustic guitar started as kinetic energy in the strings. Fortunately, it doesn't take very much acoustic energy to produce an audible sound; the kinetic energy in a string oscillating a millimeter or two does the job nicely.

The idea of low frequency and high frequency might intuitively be defined in terms of the human hearing range. Here, though, it makes sense to make the distinction based on the size of the radiator (the radiator could either a moving surface or the sound hole). This is because sound behaves differently depending on the length of the sound wave, λ, compared to the dimension, D, of the radiator. Figure 7.19 shows the propagations of high frequency sound and a low frequency sound from a radiator of fixed sized. If the wavelength is small compared to the surface dimension, $\lambda < D$, the sound is essentially directional. If the wavelength is large, $\lambda > D$, then the sound disperses in all directions with an essentially spherical wave front. If the wavelength is around the same size as the radiator, $\lambda \approx D$, then the propagation is somewhere between directional and spherical.

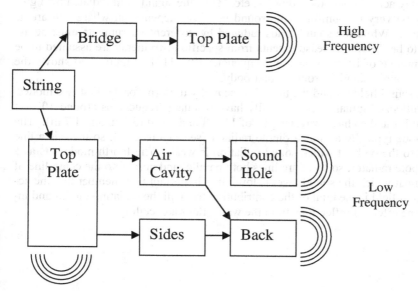

Figure 7.18 Schematic of Energy Flow in a Guitar (after Rossing [112])

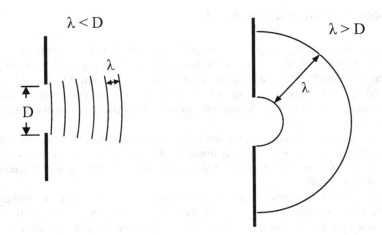

Figure 7.19 Wave Propagation as a Function of Wavelength

The wavelength is defined by the speed of sound and the frequency in Hz, $\lambda = c/f$. The speed of sound in air at sea level is approximately 1120 ft/s (341 m/s). Thus the wavelength of low E ($f = 82.4$ Hz) is 13.6 ft (4.14 m). The frequency of E played at the 12th fret of the 1st string is 659 Hz, so $\lambda = 20.4$ in (518 mm).

Recall also that, at different resonant frequencies, the radiation surfaces of a guitar may act as monopoles, dipoles, etc. Thus the sound field radiated by a guitar may be very non-uniform; the sound you hear depends on where you are as you listen. While the sound fields radiated by different instruments must be assumed to be different, measurements from specific instruments are assumed to be representative of behavior at low frequencies [25, 113]. Figure 7.20 shows the nominal sound radiation from a guitar body.

The sound field around the body is essentially uniform for the first two modes. For a full-sized guitar, these typically have resonant frequencies around 100 Hz and 200 Hz and so have wavelengths of 11.2 ft and 5.6 ft (3.4 m and 1.7 m). The third mode typically acts as a dipole radiator (see Figure 2.13), so there is a line normal to the body at which sound radiation is weak. The fourth mode acts as a quadrupole radiator, so there are lines both parallel and normal to the centerline of the body at which the sound level is low. It is important to remember that the actual sound field is the total of the contribution from all the radiating modes and includes the effect of reflections from the walls, floor and ceiling.

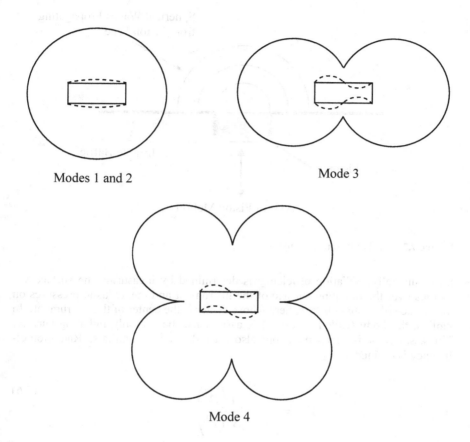

Figure 7.20 Nominal Radiation Field around a Guitar Body by Mode

A promising idea that has emerged from violin testing is that of correlating radiation efficiency and radiation damping with subjective sound quality. Results reported by Bissinger [114] suggest that there is an unambiguous relationship between them. Radiation efficiency, R_{eff}, relates the sound radiation from each normal mode (eigenvector) to that of a baffled piston with the same surface area and the same mean square velocity [115]. A baffled piston is the notional ideal sound radiator and is shown conceptually in Figure 7.21. A close physical equivalent is a speaker mounted on a rigid plate. Theoretically, the plate is assumed to be infinite; in practice, the edges of the plate must be more than ¼ the wavelength of the lowest frequency of interest from the piston (speaker cone).

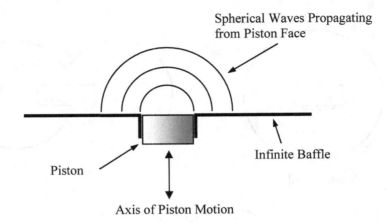

Figure 7.21 Baffled Piston Radiator

Experimentally, radiation efficiency is determined by measuring the surface ve-
locities over the radiating surface of the instrument and the acoustic pressures on
the surface of an imaginary sphere whose center is the center of the instrument. In
violins, the body (called a corpus) is assumed to be the only radiating surface.
This assumption is a reasonable one also when dealing with guitars. Radiation efficiency is defined as

$$R_{\text{eff}} = \frac{1}{\rho^2 c^2} \frac{A}{S} \frac{\langle p^2 \rangle}{\langle v^2 \rangle} \tag{7.6}$$

Where ρ is air density, c is the speed of sound, A is the area of the microphone
measurement sphere, S is the radiating area of the instrument, $\langle p^2 \rangle$ is the mean
square pressure over the microphone sphere and $\langle v^2 \rangle$ is the mean square over the
radiating surface.

Typically, R_{eff} is low when the wavelength of the radiated sound is much larger
than the radiating surface; for intuitive confirmation, note that woofers are always
much larger than tweeters. R_{eff} generally increases with increasing frequency until
the critical frequency, f_c, is reached. Bending waves propagate through a structure
at a speed that is a function of frequency – high frequency bending waves travel
faster than low frequency ones. Critical frequency is that for which bending
waves travel at the speed of sound in the material.

It is worth noting that the literature on wave propagation in structures refers to dispersive and non-dispersive waves [26]. Dispersive waves are bending waves whose propagation speed increases with increasing frequency. Thus, waves of differing frequencies being generated by a single source will disperse with respect to one another as they propagate. Non-dispersive waves are compression waves and propagate at a speed independent of the frequency. For example, sound is a non-dispersive wave and the definition of the speed of sound in an ideal gas is

$$c = \sqrt{\gamma R T} \qquad (7.7)$$

Where γ is the heat capacity ratio – also called the adiabatic index, R is the ideal gas constant and T is temperature. For air $\gamma = 1.4$. The definition has no terms for frequency or wavelength. Compressive waves in a solid are non-dispersive; the speed of sound in a solid (the velocity of a compressive wave) is given by

$$c = \sqrt{\frac{E}{\rho}} \qquad (7.8)$$

Where E is elastic modulus and ρ is the material density. Shear waves in solids are also non-dispersive, but are beyond the scope of this discussion.

Radiation damping, ζ_{rad}, is a description of energy loss due to radiation. Higher radiation damping suggests that a larger fraction of the kinetic energy in the structure is being converted to sound energy. Material damping, on the other hand, converts kinetic energy into heat. The amount of kinetic energy in a stringed instrument is very low by mechanical standards, so the increase in temperature caused by material damping is negligibly small. For example, an incandescent night light consumes about 4 W – a very small amount when energy usage is measured in kilowatt-hour. However, 4 W of acoustic power in a small room would be extremely loud. A jackhammer generates about 1 W of acoustic power.

The basic behavior of radiation efficiency and radiation damping as a function of frequency are shown in Figure 7.22. R_{eff} increases non-linearly until the f_c is reached, at which point it levels off. Radiation damping tends to increase approximately linearly until the critical frequency is reached. Above f_c, radiation damping drops in proportion to the inverse of the frequency. In violins, good sound quality has been correlated with lower critical frequency. Preliminary work on classical guitars [116, 117] suggests that increased radiation efficiency is correlated with improved sound quality.

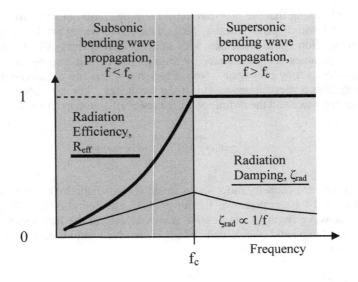

Figure 7.22 Radiation Efficiency and Radiation Damping as a Function of Frequency (after Bissinger [114])

7.4 Predicting Sound Quality

It is certainly useful to correlate structural characteristics with sound quality, but it would be even more useful to be able to predict sound quality. Then, instrument designers would be able to build in sound quality based on accurate mathematical models. However, this necessarily requires math models that capture the features of the acoustic response that are correlated with good sound quality.

Proposed methods for quantifying sound quality are based on measurements of radiated sound, so any analytical prediction method must be able to calculate at least part of the radiated sound field from predicted structural response due to string vibrations. The problem is particularly complicated because string motion, top and back motion and radiated sound are all coupled – none of them can be calculated precisely unless the effects of the others are included. At this writing, such a calculation is only possible using a discrete model with thousands of degrees of freedom. Even then, the resulting model necessarily includes a number of freely chosen parameters that must be correct in order for the resulting predictions to match experiment. These parameters include material density, material damping, stiffness and constants describing acoustic-structure coupling. Additionally, the discrete model must somehow account for the environment through which the sound propagates. For example, a finite element model must calculate pressures over the volume of interest. Furthermore, the outer edges of the discretized volume of air must be given appropriate acoustic boundary conditions. In practice, numerical models are generally compared with experimental results and then modified (sometimes called updated) so that the calculated results match the experimental ones. Design changes are then made on the validated numerical model. This, of course, undermines the entire concept of predictive analysis, but is, for now, the only practical choice.

As of this writing, reliable predictions of sound quality must wait for the combination of a robust objective description of sound quality and numerical models capable of reliably predicting the radiated sound field without first being tuned to match experimental data.

8 Guitar Electronics

The electric guitar was not a practical instrument until effective ways were developed to sense the motion of some part of the instrument and convert that motion into a proportional voltage. One of Les Paul's earliest experiments was to fix a phonograph needle to the top of an acoustic guitar [118]. Since the needle was designed to amplify vibrations caused by changes in the lateral position of the grooves in a record as it spins, it also amplified vibrations of the soundboard.

The first practical type of pickup, and the one that is still universally used on solid body electric guitars, is the electromagnetic inductive pickup. This simple device detects changes in a magnetic field caused by moving steel strings. Inductive pickups are not the only type available, though; different types of instruments require different characteristics from pickups [119].

For example, magnetic pickups are necessarily heavy and are thus difficult to install on acoustic guitars; adding weight to the soundboard reduces the amplitude of its motion in response to the vibrating strings and reduces the radiated sound level. It also has the effect of changing the resonant frequencies of the soundboard and can change the tonal quality of the sound. Piezoelectric pickups were developed as a way of amplifying acoustic guitars without adding significant weight.

In addition to the pickups themselves, a wide range of signal processing hardware and software exists to condition the output of pickups and to modify the sound produced by guitars. It makes sense, though, to focus first on inductive pickups.

8.1 Inductive Pickups

The inductive electro-magnetic pickup is a simple, elegant expression of the most basic principles of electromagnetism. The underlying idea is to set up a magnetic field and use a coil of wire to sense vibrations in that field caused by a vibrating string [120]. The result is a voltage that changes in proportion to the velocity and amplitude of the of the string vibration.

Michael Faraday discovered the relationship between electricity and magnetism that was later written in mathematical form by James Clerk Maxwell [121].

R.M. French, *Engineering the Guitar*,
DOI: 10.1007/978-0-387-74369-1_8, © Springer Science+Business Media, LLC 2009

Faraday discovered that moving a magnet through a coil of wire induced a current in the wire. He also noticed that the reverse was also true – a current was induced when the coil was moved over a stationary magnet. This effect is now called Faraday's Law and is one of Maxwell's equations.

Electromagnetic pickups are generally available in either single coil or double-coil 'humbucker' configurations. The two designs offer different frequency response and different sensitivity to electromagnetic and electrostatic noise. Figure 8.1 shows two single coil pickups mounted in an electric guitar. The metal circles are the ends of the magnets around which the coil wires are wrapped. Figure 8.2 shows top and bottom views of a humbucker pickup. There are clearly two coils and one set of magnetic poles can be adjusted to change distance with respect to the strings.

A magnet has field lines that connect the opposite poles as shown in Figure 8.3. A voltage is generated in a wire passing through the field if it moves perpendicular to the field lines (this is the basic principle that makes motors and generators work)

It would be theoretically possible to simply amplify the voltage potential across the ends of a vibrating steel string moving through the magnetic field, but it isn't practical. The voltage developed is a function of the intensity of the magnetic field and the number of turns of wire moving with respect to the field.

Figure 8.1 Single Coil Pickups

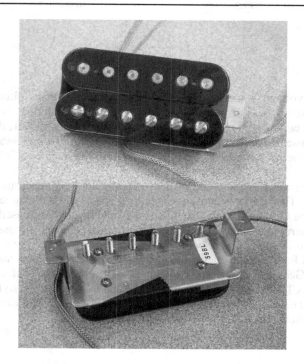

Figure 8.2 A Typical Humbucker Pickup

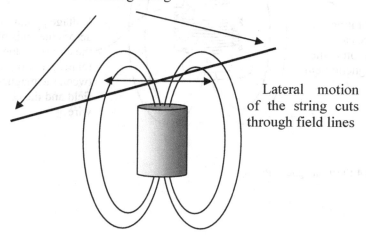

Voltage potential across the
ends of the vibrating string

Lateral motion
of the string cuts
through field lines

Figure 8.3 Wire Moving in a Magnetic Field

$$e = N \frac{\mathrm{d}\Phi}{\mathrm{d}t} \tag{8.1}$$

Where e is voltage (e stands for electromotive force, the name initially used before the unit of emf was named to Volt after Alessandro Volta), N is the number of turns in the coil of wire and Φ is the magnetic flux. Thus, if $N=1$, the wire will be relatively insensitive to changes in the magnetic field. To increase sensitivity, either the number of turns of wire or the strength of the magnetic field has to increase.

The solution was to wind a coil of fine wire around the magnet itself as shown in Figure 8.4. This works because the vibrating string and the magnetic field are coupled and the vibrating string also affects the field by making it oscillate. Equation (8.1) requires only that there be relative motion between the magnetic field and the wire coil – either one or both could be moving.

It is typical for electromagnetic pickups to have several magnets inside a single coil of wire. Generally, there are six magnets – one per string – but some designs use as many as 11. Many pickups, such as the one in Figure 8.2 also have screws attached to the magnets that act as extensions of the magnet. They can be adjusted to vary the distance from the string and, thus, alter the tone.

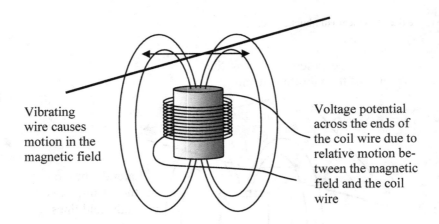

Vibrating wire causes motion in the magnetic field

Voltage potential across the ends of the coil wire due to relative motion between the magnetic field and the coil wire

Figure 8.4 Electromagnetic Pickup

Figure 8.5 Inductive Wheel Speed Sensors With and Without Protective Cover

Inductive sensors like those used on electric guitars are also used for a variety of other applications. For example the wheel speed sensors on some automotive anti-lock brake systems use inductive sensors to sense the passing rate of a toothed disk on the wheel hub to monitor wheel speed. Figure 8.5 shows two wheel speed sensors, one with the cover removed to expose the wire coil.

The most common configurations of pickups are single coil and double coil or humbucker. A humbucker typically uses a bar magnet with steel pole pieces or a steel bar at each end. The wire coil is wrapped first around one pole and then the other as shown in Figure 8.6. The coil of wire wrapped around the magnets can act as an antenna and respond to any electromagnetic noise in the environment. Not surprisingly, the most common source of this noise is a 60 Hz signal from wires carrying AC current. It manifests itself as a low hum in the sound coming from the amplifier.

The concept behind the humbucker is what would now be called common mode rejection. The two coils are in series with each other, but are wound in opposite directions. They both sense the same electromagnetic interference, but one coil senses it 180° out of phase with the other one. Since the magnetic poles are reversed, the response to the strings is not affected by the change in coil direction. The effect is to cancel out the noise while retaining the signal from the vibrating string.

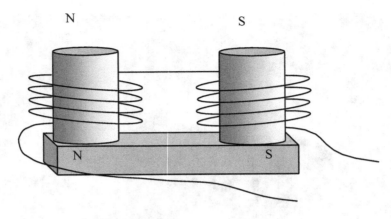

N   S

N   S

Figure 8.6 Windings on a Humbucker Pickup

Figure 8.7 shows the original patent for the Gibson humbucker pickup (Seth Lover was a Gibson employee). The sketch labeled Fig. 6 on the patent clearly shows two poles attached to a central magnet and the coils wound in opposite directions. This design became known colloquially as the Gibson PAF Humbucker with PAF standing for 'Patent Applied For'.

Humbuckers produce a sound clearly different from single coil pickups. Single coil pickups are often described as having a 'clear' or 'bright' sound while humbuckers are 'warm' or 'mellow'. The difference in tone results from the difference in the aperture – the length of the string that is sampled by the pickup- and from the fact that the two coils give the circuit two closely-spaced resonant frequencies.

The concept of aperture is a natural extension of the observability of mode shapes as shown in Figures 4.3 and 4.4. An ideal sensor that measures the velocity of string at a single point would observe a clearly defined combination of the frequencies. However, sensing the motion over a short length of the string (this length being the pickup's aperture) has the effect of averaging the signal, changing its frequency content.

As an example, consider the simple case of a string whose motion consists of components from only the first three modes. The string is tuned to a fundamental frequency of 196 Hz (open G string). The frequency content of the string is normalized so that the maximum displacement of mode 1 is 1, the maximum displacement of mode 2 is ½ and the maximum displacement of mode 3 is 1/3. The expression for the displacement at any point on the string at any time is then

$$y(x,t) = \sin(\pi x)\sin(2\pi f_0 t) + \frac{1}{2}\sin(2\pi x)\sin(4\pi f_0 t) + \frac{1}{3}\sin(3\pi x)\sin(6\pi f_0 t) \tag{8.2}$$

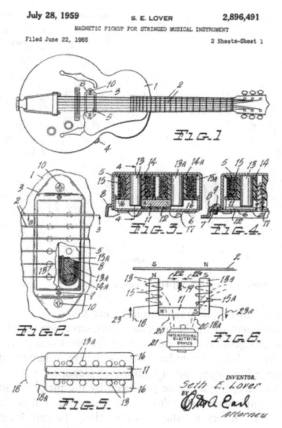

Figure 8.7 Patent for the Humbucker Pickup

Assume a pickup is placed 85% of the distance from the nut to the saddle. The string motion as a function of time at that position is

$$y(t) = 0.4540\sin(1232t) - 0.4045\sin(2463t) + 0.3292\sin(3694t) \qquad (8.3)$$

However, a humbucker pickup has two coils whose centers are about 0.75 in (19 mm) apart. If the first coil was aligned with the 85% point on the string, then the second coil would be aligned with the 88% point. The string position there as a function of time is

$$y(t) = 0.3387\sin(1232t) - 0.3187\sin(2463t) + 0.2869\sin(3694t) \qquad (8.4)$$

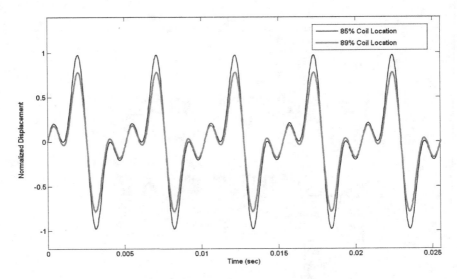

Figure 8.8 String Motion at Locations of Humbucker Coils

The two signals are shown in time domain in Figure 8.8.

Since the development of the humbucker, there have been numerous attempts to improve the noise rejection characteristics of the electromagnetic pickup without affecting the tone. While some have been effective, a large majority of electric guitars still use either single coil or humbucker pickups.

One other factor affecting the tone produced by the pickup is its electromagnetic impedance. Impedance is analogous to resistance, but differs in being a function of frequency. In fact, resistance is essentially the zero frequency component of impedance. Unfortunately, there is no convenient impedance meter like there is for resistance. Thus, pickup manufacturers generally supply impedance values for their products.

A single coil pickup typically uses thin gauge wire wound around a bobbin formed by pressing cylindrical magnets into two flat end caps, colloquially called flatwork, made of vulcanized fiber. The wire is then wound directly onto the rod magnets. Figure 8.9 shows the flatwork and magnets for a typical single coil pickup.

A very compact, non-standard, single coil pickup is shown in Figure 8.10. It has a relatively small number of turns on the coil and is intended to be used with a pre-amp mounted in the instrument.

Figure 8.9 Components of a Single Coil Pickup (Image courtesy of Stewart-MacDonald, www.stewmac.com)

Figure 8.10 A Compact Single Coil Pickup (image the author, reproduced courtesy of Taylor Guitars, www.taylorguitars.com)

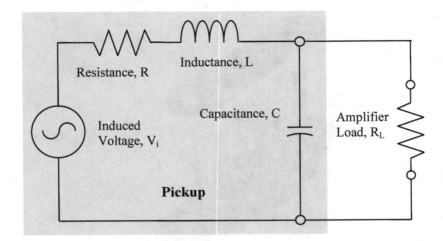

Figure 8.11 Equivalent RLC Circuit for a Single Coil Pickup

A simple circuit diagram for a single coil pickup is shown in Figure 8.11. The resistance in the coil is a function of the wire gauge (diameter) and the number of turns (length of the wire). Fender gives the resistance for a '57 single coil strat pickup as 5.6 kΩ. The inductance is 3.0 H (henry, H, is the unit for inductance) [122]. For simplicity, this diagram omits any controls on the instrument that are between the pickup and amplifier. Instrument controls typically use 250 kΩ or 500 kΩ potentiometers (variable resistors) and one or more capacitors. Depending on the setting the player chooses, these can add a significant resistance between the pickup and the amp.

A typical pickup may have 6500 turns of wire in the coil, so the inductance is relatively high compared to examples shown in textbooks. The coil wire is very fine and may be in excess of 2500 feet (762 m) of wire in a single pickup. This is reflected in the high resistance and also by a capacitance in parallel with the resistance and inductance.

This is a simple RLC circuit that can be analyzed using methods from a basic electronics class. This equivalent circuit includes components in both series and parallel. To analyze the series-parallel circuit, it can be replaced with an equivalent series circuit. The two parallel components are replaced with a single equivalent series component of identical impedance.

Electrical impedance is essentially resistance to time dependent (AC) voltage and is denoted as Z. Because impedance includes phase lag, expressions for individual components are usually expressed in complex notation. Impedance expressions for resistors, inductors and capacitors are

$$Z_R = R$$
$$Z_L = j\omega L$$
$$Z_C = \frac{-j}{\omega C} \tag{8.5}$$

Where $j = \sqrt{-1}$, rather than i. This is done so that i can be reserved to represent current.

The components are assigned impedance values as shown in Figure 8.12. Where

$$Z_1 = Z_R + Z_L = R + j\omega L \tag{8.6}$$

Z_2 and Z_3 are replaced with an equivalent impedance, Z_{23} as shown in Figure 8.13.

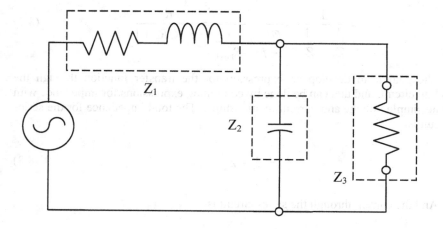

Figure 8.12 Equivalent RLC Circuit for a Single Coil Pickup

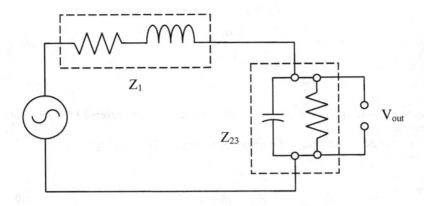

Figure 8.13 Equivalent Circuit with Two Impedance Values

$$Z_1 = R + j\omega L$$

$$Z_{23} = \cfrac{1}{\cfrac{1}{Z_2} + \cfrac{1}{Z_3}} = \cfrac{1}{\cfrac{\omega C}{-j} + \cfrac{1}{R_{\text{Load}}}} = \cfrac{R_{\text{L}}}{j\omega CR_{\text{L}} + 1} \qquad (8.7)$$

The goal is to develop an expression for the transfer function through the pickup circuit, and this can be done by combining expressions for impedance with some simple voltage and current relationships. The total impedance for the series circuit is

$$Z_{\text{T}} = Z_1 + Z_{23} \qquad (8.8)$$

And the current through the series circuit is

$$I = \frac{V_{\text{in}}}{Z_{\text{T}}} = \frac{V_{\text{in}}}{Z_1 + Z_{23}} \qquad (8.9)$$

Finally, the output voltage is

$$V_{\text{out}} = I Z_{23} \quad or \quad I = \frac{V_{\text{out}}}{Z_{23}} \qquad (8.10)$$

Now, the transfer function is easy to write out explicitly

$$H = \frac{V_{out}}{V_{in}} = \frac{Z_{23}}{Z_1 + Z_{23}}$$

$$= \frac{\dfrac{R_L}{j\omega CR_L + 1}}{R + j\omega L + \dfrac{R_L}{j\omega CR_L + 1}} \tag{8.11}$$

$$= \frac{R_L}{(R + j\omega L)(j\omega CR_L + 1) + R_L}$$

As an example, consider an EMG model SEHG dual coil (humbucker) pickup [123]. When using only one coil, R=4350 Ω and L=2.32 H. The resonant frequency is 3700 Hz and the impedance at 3700 Hz is 80.8 kΩ. Assume this pickup is plugged into an amp with an input resistance of 100 kΩ (e.g. a Fender model 212R amp). The capacitance is not provided by the manufacturer. Choosing C = 652 pF gives a resonant frequency of 3702 Hz and Z at resonance of 99.9 kΩ. The calculated transfer function is shown in Figure 8.14.

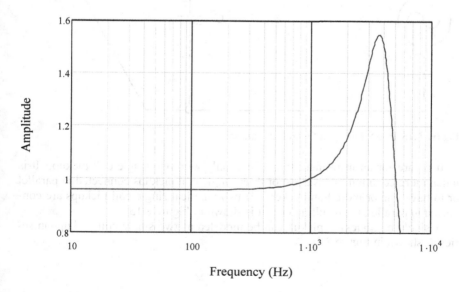

Figure 8.14 Transfer Function of Single Coil Pickup

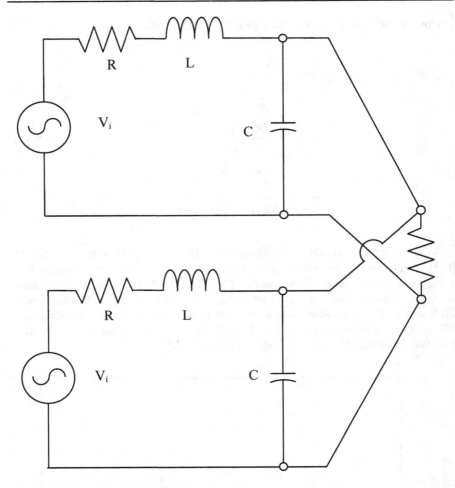

Figure 8.15 Two Single Coil Pickups in Parallel

It is rare for an instrument to be fitted only with one single coil pickup. It is much more common to have two or three single coil pickups connected in parallel or to have one or more humbuckers. If two identical single coil pickups are connected in parallel, the resulting circuit is shown in Figure 8.15.

A single humbucking pickup can be modeled as two single coil pickups in series as shown in Figure 8.16.

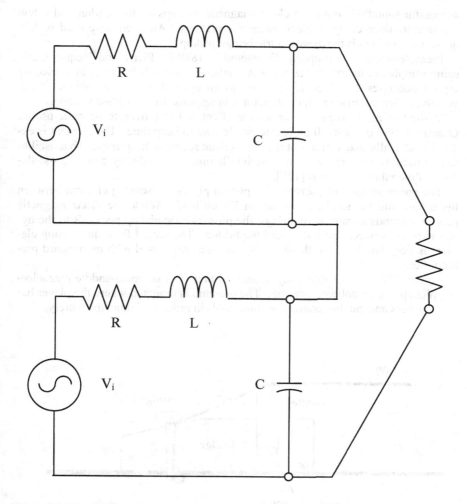

Figure 8.16 Equivalent Circuit for a Humbucking Pickup

The other circuit elements present in almost every electric guitar are tone and volume controls. Both are based on potentiometers (variable resistors), usually with a maximum resistance of either 250 kΩ or 500 kΩ.

8.2 Piezoelectric Pickups

While electromagnetic pickups are nearly universally used on solid body guitars, a larger range of pickups is installed on acoustic guitars. There are a number of successful soundhole-mounted electromagnetic pickups on the market and a few instruments have compact electromagnetic pickups. An alternative used widely on acoustic guitars is the piezoelectric bridge pickup.

Piezoelectricity is a property discovered in 1880 by Pierre and Jacques Curie using samples of quartz, topaz and a few other compounds [124]. A piezoelectric crystal generates a voltage in response to an applied mechanical stress. Conversely, a piezoelectric crystal will deform in response to an applied voltage.

While this might seem like an arcane effect, it has proven to be quite useful. Quartz watches use a small piezoelectric element to keep time. The element is excited electrically and vibrates at a characteristic resonant frequency. That oscillation is used as a reference so the watch tells time essentially by counting oscillations of the vibrating crystal [125].

The common form of piezoelectric pickup places the sensing element between the bridge and the saddle as shown in Figure 8.17. While the electromagnetic pickup responds to motion of strings, the piezoelectric pickup responds to the dynamic force between the saddle and the bridge. The output from the sensing element is very low level, so these pickups are generally used with an onboard preamplifier.

Figure 8.18 shows a thin body acoustic guitar with an under-saddle piezoelectric pickup and an onboard preamp. The pre-amp is mounted where the player has ready access and can be removed without tools in order to change the battery.

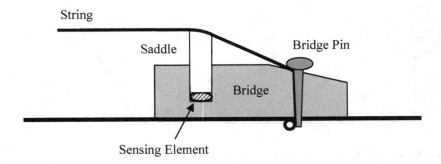

Figure 8.17 Location of Piezoelectric Sensing Element

Preamp

Figure 8.18 Thin Body Acoustic Electric Guitar with Under-Saddle Pickup

8.3 Seismic Pickups

A type of pickup that is becoming more popular is the seismic pickup – really an accelerometer permanently mounted on the soundboard. An accelerometer senses acceleration by monitoring the relative motion of a small seismic mass as the entire device experiences acceleration [126]. There are a number of different types

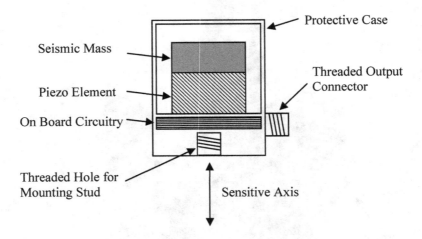

Figure 8.19 Components of a Typical Piezoelectric Accelerometer

of accelerometer on the market, though the ones used for structural testing are usually piezoelectric. Figure 8.19 shows the basic components.

These accelerometers are available in many different configurations and some miniature units weigh less than a gram. Frequency response typically ranges from a few Hz at the lower end to around 10 kHz on the upper end and vary with the size an configuration of the accelerometer. Fixing one of these devices to the soundboard of an acoustic guitar and sending the output to an amplifier shows that they can make very good pickups.

Conceptually, a soundboard mounted accelerometer differs from an electromagnetic pickup or an under saddle piezoelectric pickup in an important way – the resulting signal is proportional to motion (acceleration) of the soundboard. This is also the mechanism that radiates sound. While sound radiation is a function of the surface velocity rather than acceleration, the two are separated only by a derivative.

$$x = e^{i\omega t} \qquad\qquad x = \sin \omega t$$

$$\dot{x} = v = i\omega\, e^{i\omega t} \quad \text{or} \quad \dot{x} = v = \omega \cos \omega t \qquad\qquad (8.12)$$

$$\ddot{x} = a = -\omega^2\, e^{i\omega t} \qquad \ddot{x} = a = -\omega^2 \sin \omega t$$

Instrumentation accelerometers are designed to have a flat frequency response across their entire dynamic range and are quite expensive compared to a conventional inductive pickup. However, some manufacturers have developed seismic pickups that are well-suited for use in guitars. One example is the model DPU-3

Figure 8.20 A Piezoelectric Accelerometer and Pre-Amplifier (image courtesy Duncan Turner Acoustic Research, www.d-tar.com)

by Duncan Turner Acoustic Research as shown in Figure 8.20. This device uses a piezo film sensing element rather than a piezo ceramic element as used in laboratory accelerometers and some older contact pickups.

8.4 Pre-Amplifiers

Electric guitars fitted only with inductive pickups generally don't have on-board pre-amplifiers, but acoustic and hybrid instruments fitted with piezoelectric pickups or a combination of pickups generally do. There are a several possible reasons for wanting to use a pre-amp, but a primary one is to avoid an impedance mismatch between in the pickup and the amp [127].

Consider the case of an acoustic guitar fitted with a simple piezoelectric pickup. Unless the amplifier is specifically designed to accept an input signal from a piezoelectric pickup, it may not be able to generate a loud output. Solid state devices transfer voltage (rather than power) and there can be large voltage drops under the wrong circumstances. Impedance is the dynamic equivalent of resistance, but the basic concept can be described in terms of resistance. Impedance is a complex value that can vary with frequency, so the frequency response function of real dynamic circuits is generally not flat – the resistor analogy used here

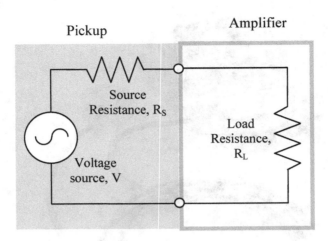

Figure 8.21 A Voltage Source with Internal Resistance Driving a Load

shouldn't be pushed too far. Figure 8.21 shows a voltage source driving a load that represents the pickup and amplifier.

If $R_S = R_L$, the two resistors would form a voltage divider and the voltage delivered to the load would be half that generated. It is worth noting that the power delivered to the load is a maximum when the resistances (impedances) are equal. The goal, however, is to deliver the maximum voltage possible to the load, thus the R_L should be much larger than R_S. The impedance of inductive pickups is generally in the range of 1-12 kΩ; this is sometimes labeled output impedance in electronics texts. The input impedance of typical guitar amplifiers is much higher than this range and values in the 1 MΩ range are not uncommon.

The output impedance of a piezoelectric pickup is necessarily high since the active element is non-conductive. Thus the output impedance is very high and the corresponding volume would be low if it were to be connected to a pre-amp designed for inductive pickups. Figure 8.22 shows an inexpensive piezoelectric contact pickup. This one, a Schatten Designs Model EP-01, is a compact device that can be mounted easily on a soundboard. Devices like this are sometimes colloquially known as 'Hot Dots'. They are useful for student instruments because of their low cost and ease of installation. The same manufacturer also offers a small onboard pre-amp with an input impedance, Z_{in}, of 10 MΩ and an output impedance, Z_{out}, of 5 kΩ, making it easily compatible with any guitar amplifier. It is also then possible to wire the piezoelectric pickup and the preamp in parallel with a lower impedance inductive pickup.

Figure 8.22 A Small Surface-Mount Piezoelectric Pickup

Combining outputs from both an inductive pickup and a piezoelectric pickup necessarily requires an impedance matching circuit. In support of a project to design a new guitar in my lab at Purdue University (www.metalsound.org), Prof. Denton Dailey of Butler County Community College has designed a preamp with inputs for both an inductive and a piezoelectric pickup. The instrument is intended for undergraduate classes and high school workshops, so low cost was a key requirement. The intended pickups are a Schatten Design EP-01 (piezo) and an EMG Select (inductive). The circuit diagram is shown in Figure 8.23.

This circuit is designed to linearly mix and continuously pan between two guitar pickups, one magnetic and one piezoelectric. The exact type of op amp used is not critical; the circuit has been prototyped using LM358 and TL071 type op amps with no noticeable difference in performance.

The gain of the individual pickup channels may be adjusted as necessary by varying the values of R_7 for the magnetic pickup and R_8 for the piezo pickup. For example, in one application, $R_7 = 100$ kΩ and $R_8 = 470$ kΩ. If additional overall gain is desired R_{16} may be increased in value, or decreased in value for less overall gain. The lower corner frequency of each of the pickup amplifiers is approximately 32 Hz. The upper frequency limit of the circuit extends well beyond the maximum range of human hearing and is not critical in this application.

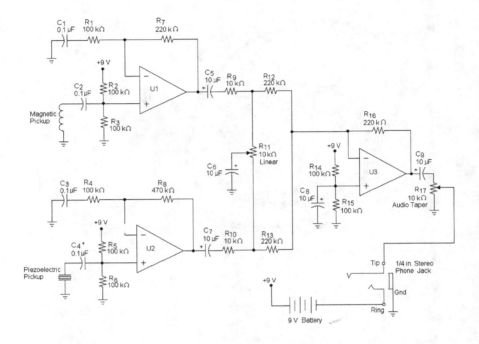

Figure 8.23 Onboard Preamp for Piezoelectric and Inductive Pickup (image courtesy, Denton Dailey, Butler County Community College)

The pickup panning network operates as follows. With potentiometer R_{11} set at the middle of its rotation, signals from both channels are applied at equal amplitudes to the input of amplifier U3, which outputs the sum of these signals. As the potentiometer is rotated, one pickup signal is attenuated while the signal from the remaining channel is passed through to the output, slightly less attenuated than with the pot centered. If the gain values for U1 and U2 are adjusted to suit the actual pickups used, the panning action of the circuit is continuous with no noticeable change in loudness, but a definite change in the tonal quality of the sound as the output is dominated by either pickup or varying combinations of both.

8.5 Grounding and Isolation

The final topic in this chapter is as much about safety as sound quality. Most electric guitar circuits, particularly those using single coil pickups are sensitive to the electromagnetic field in the space around the instrument. In the US, almost all power is transmitted as an alternating current with a frequency of 60 Hz. For historical reasons, much of the rest of the world uses 50 Hz power. In either case, there is the possibility that electromagnetic noise from AC wiring or AC-powered devices will be detected by the guitar's pickup and manifest itself as an audible hum from the amplifier with a strong 60 Hz (or 50 Hz) component.

Since 60 Hz is lower than the lowest string frequency (the low E string is generally tuned to 82.4 Hz), it would seem a simple thing to just place 60 Hz high pass filter between the guitar and the amplifier. Unfortunately, the induced signal is usually not exactly sinusoidal.   As shown in Figure 8.24 a 60 Hz sine wave with some distortion such as clipping has frequency content at integer multiples of the fundamental frequency. It becomes difficult to filter these out since a band stop filter can't tell the difference between music and noise.

There are several ways to counteract hum more directly. One obvious way is to use humbucker pickups. Another is to shield the circuitry completely; fortunately, there are some very complete descriptions available on how to do this [128]. Another method and one that is, in this author's opinion, far too popular, is to ground the circuitry to the bridge. Grounding a circuit means connecting it to a zero voltage potential so a properly shielded and grounded circuit tends not to hum – electromagnetic noise can't induce a voltage in the circuit if the shield has zero electrical potential. In colloquial terms, ground is an infinite electron bucket into which one can dump as many stray electrons as necessary.

The obvious problem is where to find a ground plane. The easiest solution is to run a ground wire to the strings and use the player as the ground plane. It is important to note that the output jack is also grounded; thus, the string ground also connects the player to the ground in the AC circuit feeding the amp power plug. This is fine as long as there is no fault in the building wiring. If there is an electrical fault, the result can be fatal.

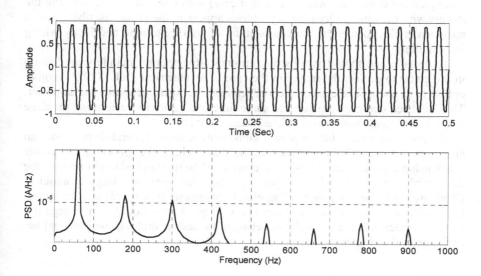

Figure 8.24 Frequency Content of a Sine Wave Clipped at +/– 90%

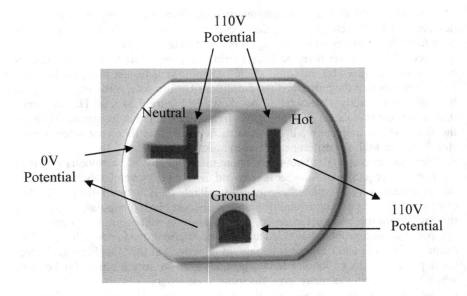

Figure 8.25 Power Outlet Showing Voltage Potentials

In the worst case, the string ground connects the player to the live wire in the power outlet. Even this is survivable as long as the player does not then ground himself or herself. Unfortunately, it is easy to grab a microphone stand (or some other piece of equipment with a grounded case) with one hand while touching the strings with the other. Occasionally, stories appear in the press describing how a musician is electrocuted by the combination of a string ground and faulty wiring in the building.

To be safe, one should avoid string grounds altogether or include some feature in the guitar to protect the player in the case of an electrical fault [129]. Another option is to use a wireless connection between the instrument and the amplifier. This has been common practice for some time with guitarist performing on stage and wireless equipment is available at modest cost [130].

If there is no option but to plug a guitar with a string ground directly into an amp with a cable, it is imperative that the player check the ground at the power outlet using a multi-meter. While the chances of being shocked might be low, the risk simply isn't worth taking, especially when there are easy ways to avoid it. Checking the ground is simple using either a voltmeter or a simple continuity tester. Figure 8.25 shows an American 110 V plug. There should be a voltage potential between hot and ground and between hot and neutral, but no potential between neutral and ground.

9 Unique Characteristics

There are several unique types of guitars and, while they all have elements in common, it is helpful to organize specific measurements and features as a reference. This chapter describes the basic features of representative guitars.

9.1 Classical Guitars

Modern classical guitars are generally similar to the designs that were common at the end of the 1800s. There is surprisingly little variation in the external appearance of classical guitars; in this sense, they have something in common with violins. This section offers dimensions and other specific characteristics of a representative classical guitar.

Classical guitars are made in many different sizes, though the dimensions vary roughly in proportion to the scale length. The dimensions given here correspond to a representative instrument with a scale length of 25.5 in (648 mm). Of course, individual instruments can vary from these measurements, but these numbers are a reasonable starting place.

Classical guitars are distinguished by several common features. They typically have flat fretboards or, less often, a large fretboard radius. They use nylon strings and have slotted headstocks with tuners mounted three on each side of the headstock. The tuners have 10 mm (0.394 in) diameter rollers spaced 35 mm (1.378 in) on center. The neck is also slightly shorter than those found on steel string acoustic guitars. The neck joins the body at the 12th fret, 12.75 in (323.8 mm) from the nut. The neck on a steel string acoustic guitar typically joins the body at the 14th fret.

Since the fretboard width varies linearly along the neck, it is convenient to define the width in terms of the distance from the nut. Figure 9.1 shows the layout of a neck based on the nut width and the string spread at the saddle. In order to fit on the page, it is not to scale.

R.M. French, *Engineering the Guitar*,
DOI: 10.1007/978-0-387-74369-1_9, © Springer Science+Business Media, LLC 2009

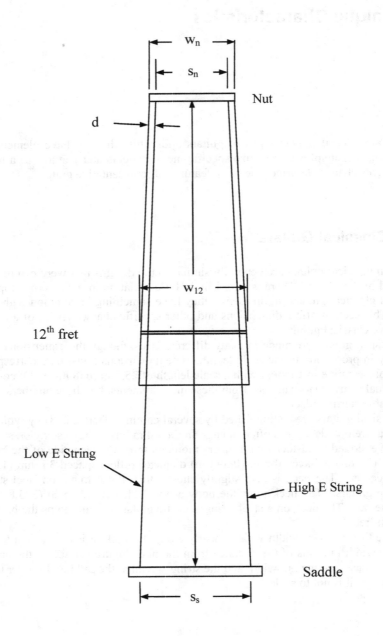

Figure 9.1 Neck Width and String Spacing for Classical Guitar

The width of the fretboard (and therefore the neck) is defined by the string spacing at the nut, s_n, and the saddle, s_s, and the distance between the centers of the outer strings and the edge of the fretboard, d. Traditionally, though, these numbers are not always defined explicitly. It is typical to define nut width, w_n, and string spacing at the saddle. d is typically assumed to be 1/8 in (3.2 mm). The string spacing at the nut is $s_n = w_n - 2d$. The nut width typically varies from 2.0 inches to 2.125 in (50.8 mm–54.0 mm). The string spacing at the saddle is typically about 2.35 in (59.6 mm).

It is convenient to define the width of the neck at the 12th fret since this is where the neck crosses the body. Width at the 12 fret is expressed as

$$w_{12} = \frac{s_n + s_s}{2} + 2d \tag{9.1}$$

And the width of the neck as a function of the distance from the nut is

$$w(x) = w_n + \frac{s_s - (w_n - 2d)}{L} = w_n + \frac{2(w_{12} - w_n)x}{L} \tag{9.2}$$

For the a neck with a 2.125 in (54.0 mm) nut, the neck width is

$$w = 2.125 + 0.019x \quad \text{(inches)}$$
$$w = 53.975 + 0.019x \quad \text{(mm)} \tag{9.3}$$

A key element of making instruments is what is often called the action. This includes the height of the strings from the fretboard, flatness of the fretboard and, where applicable, the angle between the neck and the body. Setup also includes making sure the strings do not buzz by hitting the frets as they vibrate.

The most important measurement when setting up instruments is the distance from the bottom of the strings to the top of the frets. This is generally different from string to string and also increases down the fretboard. Typically, heights are given at the first fret and the 12th fret. String height is a very personal thing, so there is no one right answer. Typical string heights are presented in Table 9.1. Relief can be measured by stopping the string at the 1st and 12th frets and finding the maximum gap between the bottom of the string and the top of the frets [131].

Table 9.1 String Heights for a Classical Guitar

	String 6 – Bass E	String 1 – Treble E
1st Fret	0.030 in (0.76 mm)	0.024 in (0.61 mm)
12th Fret	0.156 in (3.96 mm)	0.125 in (3.18 mm)
Relief	0.002 in (0.05 mm)	at the 8th fret

When cutting neck and headstock blanks, it is important to realize that there are many ways to assemble the two; three of the most common are described here. The angle between the headstock and the neck is generally between 13° and 15°. The assembly can be simply sawn from a large block, or a scarf joint can be used to attach the headstock to the neck. The band sawn neck shown in Figure 9.2a is inferior since it requires more material and produces a weaker head stock. The scarf joint shown in Figure 9.2b is traditional for classical guitars. It has the advantages of being stronger and requiring less wood to make. It is important to recall that glue is only of use in shear and the low angle of the scarf joint results in a large shearing force across the joint and a low tensile force. The glue seam on the front of the headstock is generally hidden by a decorative veneer sheet.

The author prefers the joint shown in Figure 9.2c for several reasons. It retains the strength of a scarf joint, but adds a second glued surface in shear when the fretboard is added as shown in Figure 9.3. It also requires smaller pieces of wood – a major concern since the cost of instrument grade wood is rising quickly. The author regularly makes straight, strong, attractive necks using wood salvaged from fallen logs and firewood piles. Figure 9.4 shows a neck made with this joint.

The tuners used in classical guitars are distinctly different from those used for steel stringed instruments. They are almost always mounted with three tuning machines on a single plate and are provided in matched sets so that one tuner assembly is fitted to each side of the headstock. As with nearly all tuning machines, tension in the strings is adjusted by turning a knob that rotates the roller (around which the string is wrapped) through a worm gear. Classical tuning machines nearly always have exposed gears as shown in Figures 9.4 and 9.5.

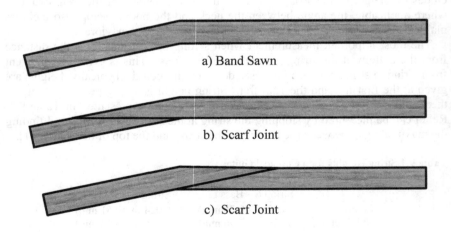

a) Band Sawn

b) Scarf Joint

c) Scarf Joint

Figure 9.2 Head and Neck Assembly

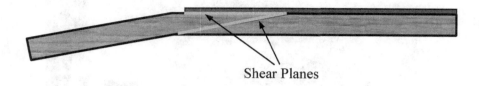

Shear Planes

Figure 9.3 Scarf Joint with Two Shear Planes

Figure 9.4 A Scarf Joint Between the Neck and Head Stock (Type C)

Figure 9.5 Tuning Machines in a Classical Guitar

All guitars require some sort of adjustment to the position of the bridge (intonation) to keep the strings approximately in tune when fretted. The intonation required for classical guitars is small because the elastic modulus of the nylon strings is low compared to steel. Different strings theoretically require different corrections, but the differences are so small when using nylon strings that a straight saddle is acceptable. Using a straight saddle necessarily requires selecting a mounting location that approximately corrects all six strings. For the design presented here, the intonation correction is 0.10 in (2.54 mm).

Classical guitars are as close to being standardized as any common style of guitar. While there is certainly no one standard classical guitar, a set of nominal or representative dimensions can be assembled. Table 9.2 presents these dimensions [5, 44, 131].

Table 9.2 Representative Dimensions for a Classical Guitar

Dimension	Value – inches	Value – mm
Scale Length	25.5	648
Headstock Width – Top	2.89	73.4
Headstock Width – Bottom	2.47	62.7
Headstock Thickness	7/8	22.2
Nut Width	2 1/8	54.0
Fretboard Thickness	1/4	6.35
Neck Width at 12th Fret	2.71	68.7
Neck Thickness at Nut (excluding the fretboard thickness)	11/16	17.5
Body Length	19.25	489
Soundhole Diameter	3 ½	88.9
Soundhole Center Location	18.68 from nut 5.93 from 12th fret	474.5 from nut
Body Depth @ Tail Block	3 ¾	95.3
Body Depth @ Neck Block	3 ½	88.9
Upper Bout Width	10.62	269.7
Width at Waist	9.00	228.6
Lower Bout Width	14.20	360.7
Nominal Soundboard Thickness	0.100	2.5
Side Thickness	0.080	2.0
Back Thickness	0.095	2.4
Tuner Spacing	1.378	35
String Spacing at Nut	1.81	46.0
String Spacing at Saddle	2.36	59.9

The development of a class on stringed instrument manufacturing and testing in the Department of Mechanical Engineering Technology at Purdue University has resulted in relatively complete geometric model for a classical guitar. This instrument is made by undergraduate students as part of the class and they have been charged with developing construction drawings. These are available at the lab web site (www.metalsound.org) and are updated as necessary. Portions of the drawings are reproduced here.

This instrument is representative of classical guitars and is intended to be simple to make for students who may have no experience with woodworking. It has no decorative elements like binding or a soundhole rosette and the bracing is probably a bit heavy. That said, the design is a forgiving one and most of the students' instruments are quite playable. Building plans for much more refined designs are available from the Guild of American Luthiers (www.luth.org).

This design has been used for experiments with alternate top materials and alternate bracing schemes. One class used braces cut from furniture grade plywood using a CNC router. While this certainly violates many accepted guidelines for bracing guitars, the results were surprisingly good.

The neck uses the scarf joint shown in Figures 9.2 and 9.3. Complete dimensions for the headstock are shown in Figure 9.6. Note that all dimensions are given in inches with dimensions in millimeters shown directly below in braces. The headstock is shown with a simple flat end, but there is no structural reason not to add some decorative shaping or inlay as shown in Figure 9.4 Indeed, many luthiers have a personal headstock shape and inlay that serves to identify their instruments.

The body shape of this instrument is the one defined in Chapter 5. The resulting dimensions are shown in Figure 9.7. The top is shown here, but the back is identical except for the soundhole and the notch for the neck tenon.

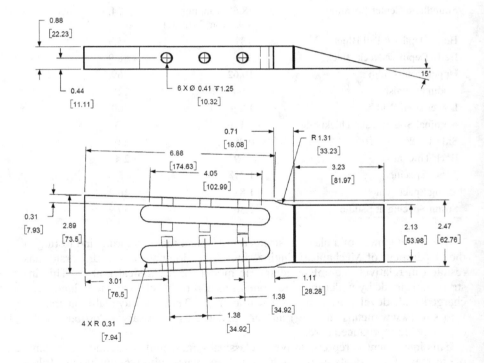

Figure 9.6 Headstock Dimensions

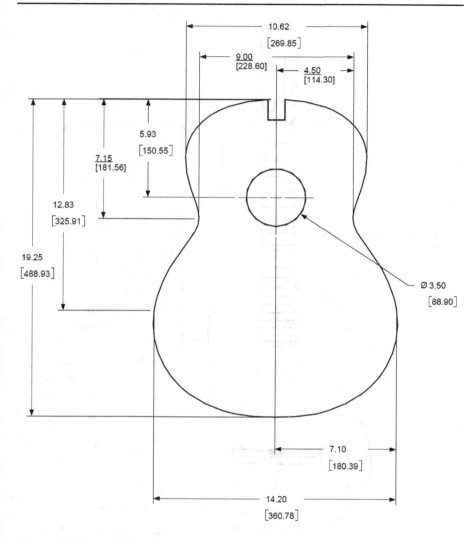

Figure 9.7 Classical Guitar Top Dimensions

Figure 9.8 shows dimensions of the complete instrument. Note that the vertical lines showing through the sound hole is a reinforcing strip over the seam at which the halves of the back plate are joined.

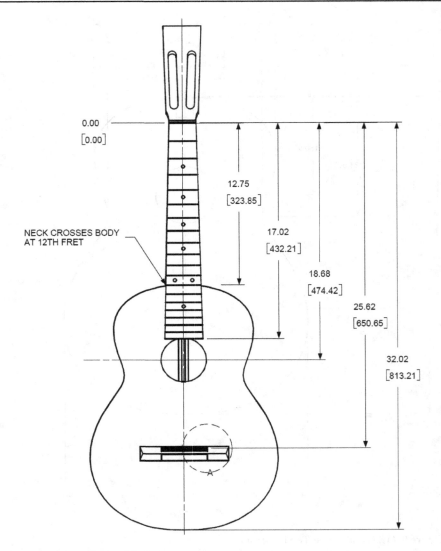

Figure 9.8 Dimensions of Complete Instrument

Figure 9.9 shows the dimensions of the blank form in which the sides are bent. The body is deeper at the bottom and this must be reflected in the shape of the unbent side. The hourglass shape of the body theoretically requires that the taper of the unbent side be a fairly complicated shape. However, making the edges parallel for part of the blank and using a straight taper to the end is very close and the difference is easily reconciled with a bit of sanding before the back is glued on.

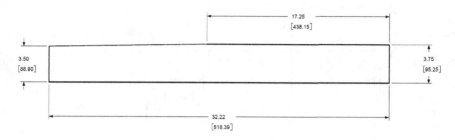

Figure 9.9 Side Dimensions

Figure 9.10 shows dimensions for the bridge. The bridge is generally made from close-grained hardwood. While ebony and rosewood are traditional, more common species such as maple and beech also work well. This part was designed to be easy for students to manufacture, but more elaborate designs are often used in commercial instruments.

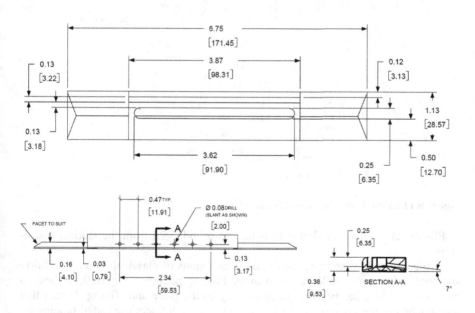

Figure 9.10 Bridge Dimensions

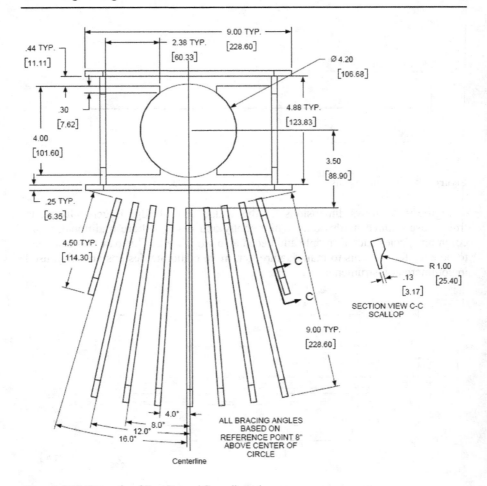

Figure 9.11 Example of Fan-Braced Soundboard

Figure 9.11 shows the dimensions for the soundboard from the student instrument.

While the body shape and the overall dimensions of classical guitars are fairly standardized, there is wide variation in the bracing patterns. Figure 9.12 shows a back with both radial bracing glued directly to the plate and 'flying' braces that are attached to the plate only at a few points. Some luthiers use radial bracing on their tops, though the flying braces seem to be largely limited to use on backs.

Figure 9.12 A Back with Radial Bracing (Photograph by Jonathan Peterson, reproduced courtesy Guild of American Luthiers, www.luth.org)

Finally the finish on classical guitars can be finer and less robust than on steel-stringed instruments. Commercially produced classical guitars often have the same type of heavy finish found on other types, but those made by individual luthiers often have particularly thin finish coats on the soundboard. French polish is often used on the soundboard while lacquer or some other more durable finish is used on the sides and back. French polishing involves applying very thin coats of shellac by hand with a small gauze or fabric pad. The resulting finish is very thin and not terribly resistant to heat and water. However, it has relatively little effect on the mechanical properties of the top and is considered to produce a better sound. Indeed, some luthiers insist that the sound quality of factory made classical guitars can often be improved by simply removing some of the finish from the soundboard.

9.2 Steel String

The vast majority of acoustic guitars sold now are fitted with steel strings. While the basic design of classical guitars is essentially standardized, steel string guitars are made in a range of shapes and sizes. Bodies usually follow one of two patterns, conventional or dreadnaught. Figures 9.13 and 9.14 show representative examples.

Figure 9.13 Conventional Body Shape, Taylor Model GA5 (Courtesy, Taylor Guitars)

Figure 9.14 Dreadnaught Body Shape (Courtesy, Taylor Guitars)

The defining difference between the steel string acoustic guitar and the nylon string classical guitar is obviously the string material, but this one difference drives most of the others. The tension required to bring steel strings to standard pitch is much higher than for nylon strings. The structure, especially the bracing, is typically heavier and the neck generally includes a truss rod. The soundboards typically use an X pattern. Figure 9.15 shows a test instrument with the bracing pattern printed onto the soundboard.

Not all the differences between steel string and classical guitars are due to structural requirements; some are driven by the different stylistic demands of the players. The necks are generally narrower and the neck typically joins the body at the 14th fret. Additionally, the fretboard is almost always radiused, though with a larger radius than for solid body electric guitars. A fretboard radius of 16 in (406 mm) is typical.

Figure 9.15 An Instrument with the Bracing Pattern Printed on the Soundboard

Because there is more variation in steel string acoustic guitars than in classical guitars, it is more difficult to present a set of representative dimensions. Table 9.3 gives values drawn from several instruments with the same scale length [132, 133, 134]. Table 9.4 presents nominal string heights for a steel string acoustic guitar Note that different players often prefer different setups and these numbers represent average values.

Table 9.3 Typical Dimensions of a Steel String Acoustic Guitar

Dimension	Value – inches	Value – mm
Headstock Width – Top	3.875	98.4
Headstock Width – Bottom	3.25	82.6
Headstock Thickness	0.600 in	15.2
Nut Width	1.75	44.5
Fretboard Thickness	0.250 (at center)	6
Fretboard Radius	16	406
Neck Width at 12th Fret	2.25	57.2
Neck Thickness at First Fret (including the fretboard thickness)	0.850	21.6
Body Length	20	508
Soundhole Diameter	4.00	102
Soundhole Center Location	5.93 (from top of body)	151
Body Depth @ Tail Block	4.40	112
Body Depth @ Neck Block	3.55	90.2
Upper Bout Width	11.75 (dreadnaught)	298
Width at Waist	10.82 (dreadnaught)	275
Lower Bout Width	16 (dreadnaught)	406
Nominal Soundboard Thickness	0.115	2.92
Side Thickness	0.100	2.54
Back Thickness	0.100	2.54
Tuner Spacing	1.325	33.7

Table 9.4 String Heights for a Steel String Acoustic Guitar

	String 6 – Bass E	String 1 – Treble E
1st Fret	0.023 in (0.58 mm)	0.013 in (0.33 mm)
12th Fret	0.090 in (2.29 mm)	0.070 in (1.78 mm)
Relief	0.002 in (0.05 mm)	at the 8th fret

9.3 Solid Body Electric Guitars

The solid body electric guitar has only been in existence since the 1940s, but is now iconic. While solid body instruments have been produced in a bewildering variety of sizes and shapes, most can probably trace their design inspiration back to three instruments: the Fender Telecaster, the Fender Stratocaster and the Gibson Les Paul.

Solid body electric guitars are much less standardized than classical guitars, so giving a list of representative dimensions is necessarily difficult. Bodies are regularly made in what seems like every shape imaginable and from a wide range of materials. However, there are some dimensions that are fairly standardized.

The most popular scale lengths are 25.5 in (648 mm), used by Fender, and 24.75 in (629 mm), used by Gibson. Instruments intended for children or smaller players have shorter scale lengths as do travel guitars designed to be easily transported and stored. Most solid body electric guitars have 21, 22 or 24 frets. Some makers (this one included) prefer 24 fret necks since that offers the player a two octave range – there are 12 semi-tone and thus 12 frets in an octave.

The method of joining the neck and body also vary, but there are three generally accepted methods: bolt on neck, glued on and neck through body. The bolt on neck is the familiar arrangement used on the Fender Stratocaster and Telecaster and adopted by many other manufacturers. It would be more correctly called a screw on neck since the fasteners are actually sheet metal screws going through the back of the body and into the back of the neck. The forces from the screws are usually distributed across the surface of the back of the body by a metal neck plate as shown in Figure 9.16.

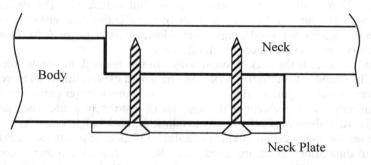

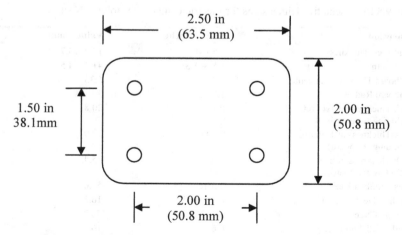

Figure 9.16 Typical Dimensions for a Bolt on Neck

The bolt on neck has several advantages. One is that the angle of the neck with respect to the body is easy to adjust. Tapered shims are readily inserted between the neck and the body to change the neck angle during setup. Additionally, the neck and body can be manufactured separately and joined only at final assembly. Mistakes made prior to final assembly affect only that part. For example, a botched neck means that only the neck is lost; the body is unaffected.

The bolt on neck design is not without some weaknesses, though. One is that loads are concentrated at the screws and wood is not the ideal material for resisting point loads. If the instrument is subjected to extremely rough treatment, it is possible for the screws to pull out of the neck. It is also possible for the neck or body to break at one of the screw holes. Also, the sustain of the instrument might be affected by the fact that the neck and body are only truly joined at the screw locations (supporting data does not, unfortunately, seem to exist). Sustain is basically the rate at which kinetic energy in the strings is dissipated. An instrument with long sustain loses energy slowly so that a note sounds for a long time.

An alternative to the bolt on neck is a glued on neck. This design is used in a number of Gibson instruments including the Les Paul and the SG. This design has the obvious advantage of distributing loads over the entire glue surface area – a much better situation for wood components. However, once the neck has been set, the geometry can be changed only with difficulty.

The third choice is the neck through body. In this method, the center block in the body is extended to form the neck. As one would expect, this is an extremely strong way to make a guitar. However, it requires a much larger piece of wood – an increasingly sticky problem as the supplies of instrument grade wood are depleted. There is also essentially no way to adjust the neck height or angle.

There are too many different kinds of solid body electric guitar for an exhaustive list of dimensions to be presented here. Rather, Table 9.5 presents nominal dimensions for a solid body guitar with a bolt on neck [133, 134, 7, 8, 10]. This is a very common configuration and may represent the majority of solid body elec-

Table 9.5 Representative Dimensions for Electric Guitar with Bolt On Neck

Dimension	Value – inches	Value – mm
Headstock Thickness	0.50–0.54	12.7–13.7
Nut Width	1 5/8–1 3/4	41.3–44.5
Fretboard Thickness (at center)	1/4	6.35
Fretboard Radius	12	305
Neck Thickness at 1st Fret (including fretboard)	0.82	20.8
Neck Thickness at 12th Fret (including fretboard)	0.90	22.9
Neck Thickness at heel (including fretboard)	1.00	25.4
Neck Width at Heel	2 3/16	55.6
Depth of Neck Pocket	0.62–0.65	16.5
Length of Neck Pocket	3.00	76.2
Depth of Pickup Pocket	0.63	16
Body Thickness	1.75	44.5
Top Thickness at Control Pocket	0.200	5.1
Scale Length	24.75–25.5	629–648

tric guitars. One measurement not presented here is the angle between the neck
and the body. The neck on many solid body guitars is parallel to the top of the
body, but this is not universal. It is not uncommon for the neck to be angled back
a few degrees. Finally Table 9.6 presents typical string heights and neck relief.

Because the range of both electronics and of body shapes is so large in solid
body electric guitars, there is no standard electronics pocket. Indeed, there is not
even agreement on whether the pocket should be on the front or back. Figure 9.17
shows Josh Hurst of Fender Guitars fitting a neck to a solid body with an elaborate
pocket in the front.

Table 9.6 String Heights for a Solid Body Electric Guitar

	String 6 – Bass E	String 1 – Treble E
1st Fret	0.024 in (0.61 mm)	0.010 in (0.25 mm)
12th Fret	0.078 in (1.98 mm)	0.063 in (1.60 mm)
Relief	0.001 in (0.03 mm)	at the 8th fret

Figure 9.17 A Solid Body with an Elaborate Front Cutout (Courtesy, Josh Hurst)

Placing the electronics pocket on the front of the body requires some sort of cover, usually a plastic pick guard. The Fender Stratocaster uses a large pick guard with a five position switch, three pickups, three potentiometers and associated wiring. Only the jack is not mounted to the pick guard. When the pick guard assembly is complete, it can be fitted into the body in a single operation. Figure 9.18 shows two pre-wired pick guards ready for installation.

Mounting the electronics from the rear of the body has the aesthetic advantage of not requiring a front cover that may obscure attractive figured wood. It is also much easier to contour or arch the front of the body.

Single coil pickups and humbucker pickups have almost standardized dimensions; there are small dimensional variations between different brands, but pickups of similar design from different vendors are generally interchangeable [135, 136]. Figure 9.19 shows the nominal dimensions for a single coil pickup and Figure 9.20 shows the nominal dimensions for a humbucker pickup.

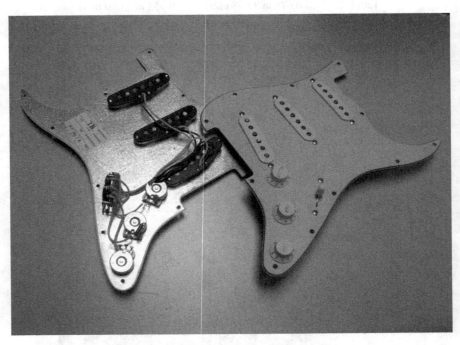

Figure 9.18 Pick Guards with Integral Electronics (Courtesy, Fender Guitars, www.fender.com)

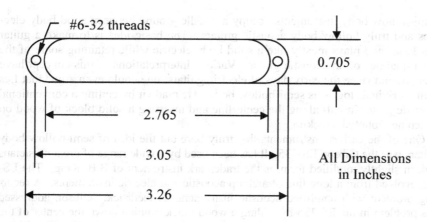

Figure 9.19 Nominal Dimensions of a Single Coil Pickup

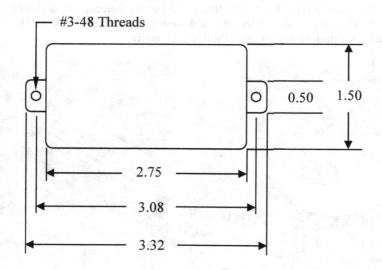

Figure 9.20 Nominal Dimensions of a Humbucker Pickup

9.4 Semi-Hollow Body Electric Guitars

Semi-hollow body instruments occupy a middle ground between solid body electrics and truly hollow body acoustic guitars. The basic idea is to make a guitar that looks and plays mostly like a solid body electric while retaining some of the tonal qualities of an acoustic guitar. Various interpretations of this concept have been around since the early days of electric guitars; one could even argue that Les Paul's original 'log' was semi-hollow body. He made it by cutting a conventional acoustic guitar in half along the centerline and inserting a solid block of wood on which he mounted a pickup.

One of the earliest instruments that truly bore out the idea of semi-hollow body guitars was the Gibson ES-335. It has been used by a wide range of great musicians and, in slightly modified form, is the trademark instrument of B.B. King. The ES-335 evolved from a long line of archtop acoustic and electric instruments. A recurring problem with amplified acoustic instruments is feedback. Gibson addressed this problem in the ES-335 by adding a wood block running down the center of the instrument and spanning the full thickness from the top to the back. The ES-335 is a thin body instrument that evolved from the ES-330, which had no center block.

There are many thin body acoustic electrics currently available. One noteworthy design is the Taylor T5 and another is the Fender Telecoustic as shown in Figure 1.15. Figure 9.21 shows an exploded view of a thin body acoustic electric developed for a student workshop at Purdue University.

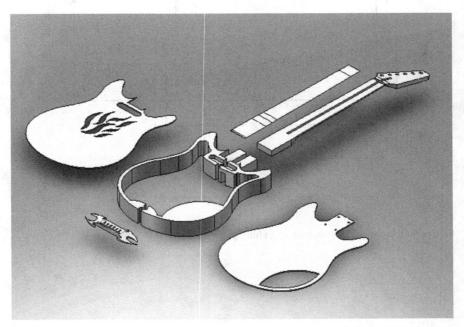

Figure 9.21 Exploded View of a Thin Body Acoustic Electric Guitar (Rendering by Eddy Efendy)

9.5 Jazz Guitars

A family of specialized instruments has evolved to meet the needs of jazz players. These typically follow one of two different patterns. The first is a large flat top instrument with an oval soundhole and a wide, thin floating bridge held in place by string tension. It was originally developed by luthier Mario Maccaferri and featured an internal resonator. It was later produced by the Selmer Company, though without the resonator. This type of instrument is often known colloquially as a Selmer or Selmer-Maccaferri guitar.

This type of instrument is most closely associated with jazz guitar great Django Reinhardt. Two representative instruments are shown in Figure 9.22. There has been a resurgence of interest in this style of guitar in the last decade or so. Even so, it represents a small fraction of the guitars being made and is very much a niche product.

The other type of instrument associated with jazz players is the archtop guitar. The typical instrument is very large with arched top and back and usually fitted with a single humbucker pickup mounted at the end of the neck. These instruments appear to have evolved from the Gibson L-5 and the typical modern example is quite refined. They are not made in large quantities (but are still more common than Selmer style guitars) and many are produced by independent luthiers. An excellent example is the Manhattan model produced by Bob Benedetto. While it is impossible to rate one maker of fine instruments as better than another, Benedetto has certainly produces instruments of the highest quality. Figure 9.23 shows a Benedetto Manhattan.

Figure 9.22 A Pair of Jazz Guitars (Image Courtesy Jingeol Lee)

Figure 9.23 The Benedetto La Venizia Archtop Jazz Guitar (Image Courtesy Bob Benedetto, www.benedettoguitars.com).

References

[1]   Paker J (1992) Music From Strings, Merlion, Wilshire UK
[2]   Sachs C (2006) The History of Musical Instruments, Dover, New York
[3]   Faber T (2004) Stradivari's Genius, Random House, New York
[4]   Chapman R (2003) Guitar: Music, History, Players, Doring Kindersley, London New York Munich Melbourne Delhi
[5]   Osborne N, ed. (2002) The Classical Guitar Book: A Complete History, Backbeat Books, San Fransisco
[6]   Bacon T (2001) The History of the American Guitar, Barnes and Noble, New York
[7]   Duchossoir AR (1981) Gibson Electrics, Hal Leonard, Milwaukee
[8]   www.lespaulonline.com
[9]   Duchossoir AR (1991) The Fender Telecaster: The Detailed Story of America's Senior Solid Body Electric Guitar, Hal Leonard, Milwakee
[10]   Denyer R, Guillory I and Crawford AM (1994) The Guitar Handbook, Knopf, New York
[11]   Caldersmith G (1995) Designing a Guitar Family. Applied Acoustics 46:3–17
[12]   Hutchins CM (1992) A 30 Year Experiment in the Acoustical and Musical Development of Violin-Family Instruments. The Journal of Acoustical Society of America 92:639–650.
[13]   Caleon I and Ramanathan S (2007) From Physics to Music: The Undervalued Legacy of Pythagoras Science and Education
[14]   Grout DJ and Palisca CV (2001) A History of Western Music, 6th Edition. W.W. Norton New York London
[15]   Hartmann WM (1997) Sound, Signals and Sensation. Springer-Verlag New York
[16]   Walker DP (1973) Some Aspects of the Musical Theory of Vincenzo Galileo and Galileo Galilei. Proceedings of the Royal Musical Association 100:33–47
[17]   Rossing TD, Moore FR and Wheeler PA (2002) The Science of Sound, 3rd ed. Pearson Education San Fransisco
[18]   Technical Committee 43 (1975) ISO 16 Acoustics – Standard Tuning Frequency (Standard Musical Pitch), International Standards Organization
[19]   Olson HF (1967) Music, Physics and Engineering. Dover Publications New York
[20]   Isacoff S (2001) Temperament. Vintage Books New York
[21]   Garland TH and Kahn CV (1995) Math and Music: Harmonious Connections. Dale Seymour Publications
[22]   Johnston I (1989) Measured Tones: The Interplay of Physics and Music. Institute of Physics London
[23]   Taylor C (1992) Exploring Music. Institute of Physics London
[24]   ISO (2007) ISO 80000–8:2007 Quantities and Units – Part 8: Acoustics. International Standards Organization Geneva
[25]   Fletcher N and Rossing TD (1998) The Physics of Musical Instruments, 2nd ed. Springer-Verlag New York

[26] Fahy F (1985) Sound and Structural Vibration. Academic Press San Diego
[27] Lai JCS and Burgess MA (1990) Radiation Efficiency of Acoustic Guitars. The Journal of Acoustical Society of America 88:1222–1227
[28] Firth IM (588) Physics of the Guitar at the Helmholtz and First Top Plate Resonances. The Journal of Acoustical Society of America 61:588–593
[29] Liessa A (1969) Vibration of Plates. NASA SP-160
[30] Howard DM and Angus J (2001) Acoustics and Psychoacoustics, 2nd ed. Focal Press Oxford
[31] Standards Secretariat, Acoustical Society of America (1986) Design Response of Weighting Networks for Acoustical Measurements ANSI S1.42-1986
[32] Fletcher H and Munson WA (1933) Loudness, Its Definition, Measurement and Calculation. The Journal of Acoustical Society of America 5:82–108
[33] ISO (2003) ISO 226:2003 Acoustics – Normal Equal-Loudness Level Contours. International Standards Organization Geneva
[34] Standards Secretariat, Acoustical Society of America (2004) ANSI S1.11-2004 Specification of Octave-Band and Fractional Octave Band Analog and Digital Filters. American National Standards Institute Washington DC
[35] ANSI (1973) American National Psychoacoustical Terminology S3.20-1973. American National Standards Institute Washington DC
[36] Cohen L (1994) Time Frequency Analysis. Prentice Hall Englewood Cliffs
[37] Cohen L (1989) Time-Frequency Distributions – A Review. Proceedings of the IEEE 77:941–981
[38] Choi HI and Williams WJ (1989) Improved Time-Frequency Representation of Multi-component Signals Using Exponential Kernels. IEEE Transaction on Acoustics, Speech, and Signal Processing 37:862–871
[39] http://www.ptc.com/products/mathcad/mathcad14/promo.htm, Web site last visited 8 Oct 2007.
[40] Mott RL (2007) Applied Strength of Materials. Prentice Hall Upper Saddle River, New Jersey.
[41] Bécache E, Chaigne A, Dervaux G and Joly P (2005) Numerical Simulation of a Guitar. Computers and Structures 83:107–126
[42] D'Addario Catalog, http://www.daddario.com/DAddarioFrettedCatalog.pdf. Last visited Jan 4, 2008.
[43] Forest Products Lab (2002) Wood Handbook – Wood as an Engineering Material. USDA Forest Service
[44] Bruné R (2004) Eight Concerns of Highly Successful Guitar Makers. American Luthierie 79:6–21
[45] Benedetto R (1994) Making an Archtop Guitar. Centerstream Publishing
[46] Kasha M and Kasha N (1982) Applied Mechanics and the Modern String Instrument – Classical Guitar. Journal of Guitar Acoustics 6:104–121
[47] Sali S and Kopac J (2002) Positioning of Braces on a Guitar Soundboard. Proceedings IMAC XX 709–715.
[48] Reynolds R (2006) Double-Top Guitars. American Luthier 88:8–12
[49] DOT/FAA/AR-02/97 (2002) Shear Stress-Strain Data for Structural Adhesives. US Department of Transportation/Federal Aviation Administration Office of Aviation research
[50] www.titebond.com
[51] ASTM D905-98 (1998) Standard Test Method for Strength Properties of Adhesive Bonds in Shear by Compressive Loading. American Society for Testing and Materials, ASTM, West Conshohocken, PA.
[52] http://en.wikipedia.org/wiki/Hide_glue

[53] Buck SL (1990) A Study of the Properties of Commercial Liquid Hide Glue and Traditional Hot Hide Glue in Response to Changes in Relative Humidity and Temperature. Proceedings, Wooden Artifacts Group

[54] http://en.wikipedia.org/wiki/Shellac

[55] http://en.wikipedia.org/wiki/Tung_oil

[56] Mottola RM (2008) Sustain and Electric Guitar Neck Type. American Lutherie 91 (in press)

[57] Barker J (2001) Violin Making: A Practical Guide. Crowood Press.

[58] Beer FP, Johnston ER and DeWolf JT (2002) Mechanics of Materials, 3rd ed. McGraw Hill.

[59] Dally JW, Riley WF and McConnell KG (1993) Instrumentation for Engineering Measurements. John Wiley and Sons.

[60] www.cites.org

[61] Kreyszig E (2005) Advanced Engineering Mathematics, 9th ed. John Wiley and Sons

[62] Inman DJ (2001) Engineering Vibration, 2nd ed. Prentice Hall

[63] Ramirez RW (1985) The FFT: Fundamentals and Concepts. Prentice Hall

[64] McConnell KG (1993) Vibration Testing: Theory and Practice. John Wiley and Sons.

[65] Campbell DM Meyers A and Greated CA (2006) Musical Instruments: History, Technology and Performance of Instruments of Western Music. Oxford University Press

[66] Galileo G (1638) Dialogues Concerning Two New Sciences. Dover Publications (reprint)

[67] Bolwell JE (1999) How Realistic is the D'Alembert Plucked String?. European Journal of Physics 20:313–320.

[68] Shankland RS and Coltman WJ (1939) Departures of the Overtone of a Vibrating Wire From a True Harmonic Series. Journal of the Acoustical Society of America 10:161–166.

[69] Byers G (1996) Classic Guitar Intonation. American Lutherie 47:368

[70] Bolwell JE (1997) The Flexible String's Neglected Term. Journal of Sound and Vibration 206:618–623.

[71] Railsback OL (1938) Scale Temperament as Applied to Piano Tuning. of the Acoustical Society of America 9:274.

[72] Stetson KA (1981) On Modal Coupling in String Instrument Bodies. Journal of Guitar Acoustics 3:23–31

[73] Cloud GL (1998) Optical Methods of Engineering Analysis. University Press.

[74] Beranek LL and Sleeper HP (1946) The Design and Construction of Anechoic Sound Chambers. Journal of the Acoustical Society of America 18:140–150.

[75] Gierlich HW (1992) The Application of Binaural Technology. Applied Acoustics 36:219–243.

[76] Rayleigh, JWS (1976) The Theory of Sound. Dover Publications.

[77] Hutchins, CM (1983) A History of Violin Research. Journal of the Acoustical Society of America 73:1421–1440.

[78] Schelling, JC (1963) The Violin as a Circuit. Journal of the Acoustical Society of America, 35:326–338.

[79] Pritchard PJ (2007) MathCAD: A Tool for Engineers and Scientists. McGraw-Hill.

[80] Logan DL (2006) First Course in the Finite Element Method, 4th ed. Nelson.

[81] Caldersmith G (1982) The Guitar Frequency Response. Journal of Guitar Acoustics, 6:1–9.

[82]   Christensen O and Vistisen BB (1980) Simple Model for Low Frequency Guitar Function. Journal of the Acoustical Society of America, 63:758–766.

[83]   Christensen O and Vistisen BB (1980) Quantitative Models for Low Frequency Guitar Function. Journal of Guitar Acoustics, 6:10–25.

[84]   French RM (2007) Structural Modification of Stringed Instruments. Mechanical Systems and Signal Processing, 21:98–107.

[85]   Habault D, ed. (1999) Fluid-Structure Interactions in Acoustics. Springer-Verlag.

[86]   Uqural AC, Fenster SK and Fenster S (2003) Advanced Strength and Applied Elasticity. Pearson Education.

[87]   Brebbia CA and Dominguez J (1992) Boundary Elements: An Introductory Course. McGraw-Hill.

[88]   Schey HM (2005) Div, Curl Grad and All That: An Informal Text on Vector Calculus. W.W. Norton and Co.

[89]   French RM (2006) A Different Way of Defining Body Shapes. American Luthierie, 88:52–57.

[90]   Olsen T and Burton C (1990) Lutherie Tools: Making Hand and Power Tools for Musical Instrument Building. Guild of American Luthiers.

[91]   French RM (2008) Response Variation in a Group of Acoustic Guitars. Sound and Vibration, 42:18–23.

[92]   Shainin RD (1993) Strategies for Technical Problem Solving. Quality Engineering, 5:433–448.

[93]   French RM, Nowland M, Hedges D and Greely D (2004) Reducing Noise from Gas Engine Fuel Injectors. International Journal of Vehicle Noise and Vibration, 1:83–96

[94]   Orduna-Bustamante F (1992) Experiments on the Relation Between Acoustical Properties and the Subjective Quality of Classical Guitars. Journal of the Catgut Acoustical Society, 2:20–23.

[95]   Jaroszewski A and Rakowski JZ (1978) Opening Transients and the Quality of Classical Guitars. Archives of Acosutics, 3:79–84.

[96]   Oribe J (1985) The Fine Guitar. Mel Bay Publications.

[97]   Bruné RE (1980) Soundboard Bracing and the Development of the Classical Guitar. Guild of American Luthiers Quarterly, 9.

[98]   Pratt RL and Doak PE (1976) A Subjective Rating Scale for Timbre. Journal of Sound and Vibration, 45:317–328.

[99]   Boehm T and Miklaszewski K (1986) Estimation of Guitar Sound Quality. Archives of Acoustics, 11:203–229.

[100]   Šali S and Kopač J (2000) Measuring the Quality of Guitar Tone. Experimental Mechanics, 40:242–247.

[101]   Moore and Glasberg (1983) Suggested Formulae for calculating auditory filter bandwidths and excitation patterns, JASA 74:750–753.

[102]   Gilchrist N and Grewin C eds (1996) Collected Papers on Digital Audio Bit-rate Reduction. The Audio Engineering Society.

[103]   Moore BCJ (1997) An Introduction to the Psychology of Hearing. Academic Press.

[104]   David HA (1988) The Method of Paired Comparisons. Oxford University Press.

[105]   Bourgeoi D (1990) Voicing the Steel String Guitar. American Lutherie, 24.

[106]   Caldersmith G (1984) Testing Tonewood Samples, reprinted in Big Red Book of American Lutherie, Vol. 2. Guild of American Luthiers.

[107]   Peterson M, Silber M, Steinegger R, Bourgeois D, Ford F and van Linge S (1996) Retro Voicing the Flat Top Guitar. American Lutherie, #47.

[108]   Williams J (1995) Lattice Bracing Guitar Tops. American Lutherie, 43.

[109]   Hutchins CM (1981) The Acoustics of Violin Plates. Scientific American, 245.

[110]   Carruth A Free Plate Tuning, Big Red Book of American Lutherie, 3:136–172

[111] Caldersmith G (1981) Physics at the Workbench of the Luthier. Journal of Guitar Acoustics, 2:28–34
[112] Rossing TD (1988) Sound Radiation from Guitars. American Lutherie, 16.
[113] Le Pichon A, Berge S and Chaigne A (1998) Comparison between Experimental and Predicted Radiation of a Guitar. Acustica united with Acta Acustica, 84:136–145.
[114] Bissinger G (2005) A Unified Materials-Normal Mode Approach to Violin Acoustics. Acta Acoustica United with Acoustica, 91:214–228.
[115] Doyle JF (1997) Wave Propagation in Structures, 2nd ed. Springer Verlag.
[116] Lai JCS and Burgess MA (1990) Radiation Efficiency of Acoustic Guitars. Journal of the Acoustical Society of America, 88:1222–1227.
[117] Boullosa R, Orduna-Bustamante F and Pérez López A (1999) Tuning Characteristics, Radiation Efficiency and Subjective Quality of a Set of Classical Guitars. Applied Acoustics, 56:183–197.
[118] Gugliotta G (2005) The Log Puts Paul in Ranks for Top Inventors. Washington Post, 5/16/05
[119] Ban FT (2007) Analysis of Electric Guitar Pickups. Proceedings of the 25th International Modal Analysis Conference, Orlando.
[120] Smith GS (1997) An Introduction to Classical Electromagnetic Radiation. Cambridge University Press
[121] Guillen M (1995) Five Equations that Changed the World. Hyperion.
[122] Schultz ME (2007) Grob's Basic Electronics, 10th ed. McGraw-Hill
[123] www.emginc.com, Web site for EMG Pickups, last visited 8 March 2008.
[124] Giancoli D (2008) Physics for Scientists and Engineers, 4th ed. Addison Wesley.
[125] Macaulay D (1998) The New Way Things Work. Dorling Kindersley Limited
[126] Figliola RS and Beasley DE (2000) Theory and Design for Mechanical Measurements. John Wiley and Sons.
[127] Beckwith TG, Marangoni RD and Lienhard JH (1993) Mechanical Measurements, 5th ed. Addison Wesley Longman.
[128] Schmitt R (2002) Electromagnetics Explained. Elsevier Science.
[129] Hosler D (2006) Musical Instrument String Ground Circuit Breaker. US Patent 7238877.
[130] Duncan B (2002) The Live Sound Manual: Getting the Best Sound at Every Gig. Backbeat Books.
[131] Sloane I (1989) Classic Guitar Construction, The Bold Strummer.
[132] www.fender.com
[133] Cumpiano WR and Natelson JD (1994) Guitarmaking: Tradition and Technology, Chronicle Books.
[134] Personal correspondence with Josh Hurst, Fender Musical Instrument Co.
[135] www.seymourduncan.com.
[136] www.stewmac.com

Index